Story Mode

Story Mode

The Creative Writer's Guide to Narrative Video Game Design

Julialicia Case, Eric Freeze, and Salvatore Pane

BLOOMSBURY ACADEMIC
LONDON • NEW YORK • OXFORD • NEW DELHI • SYDNEY

BLOOMSBURY ACADEMIC
Bloomsbury Publishing Plc
50 Bedford Square, London, WC1B 3DP, UK
1385 Broadway, New York, NY 10018, USA
29 Earlsfort Terrace, Dublin 2, Ireland

BLOOMSBURY, BLOOMSBURY ACADEMIC and the Diana logo are trademarks of
Bloomsbury Publishing Plc

First published in Great Britain 2024

Cover design by Namkwan Cho
Cover image © Shutterstock

A catalogue record for this book is available from the British Library.

Library of Congress Cataloging-in-Publication Data

Names: Case, Julialicia, author. | Freeze, Eric, 1974- author. | Pane, Salvatore, author.
Title: Story mode : the creative writer's guide to narrative video game design / Julialicia
Case, Eric Freeze and Salvatore Pane.
Description: London ; New York : Bloomsbury Academic, 2024. |
Includes bibliographical references.
Identifiers: LCCN 2023030741 (print) | LCCN 2023030742 (ebook) | ISBN 9781350301368
(hardback) | ISBN 9781350301375 (paperback) | ISBN 9781350301382 (adobe pdf) |
ISBN 9781350301399 (epub)
Subjects: LCSH: Video games–Authorship. | Video games–Design.
Classification: LCC GV1469.34.A97 C37 2024 (print) | LCC GV1469.34.A97 (ebook) |
DDC 794.8/3–dc23/eng/20231018
LC record available at https://lccn.loc.gov/2023030741
LC ebook record available at https://lccn.loc.gov/2023030742

ISBN: HB: 978-1-3503-0136-8
PB: 978-1-3503-0137-5
ePDF: 978-1-3503-0138-2
eBook: 978-1-3503-0139-9

Typeset by Deanta Global Publishing Services, Chennai, India
Printed and bound in Great Britain

To find out more about our authors and books visit www.bloomsbury.com and
sign up for our newsletters.

Contents

Part III Practical Tools and Approaches 179

12 Tool Suites and the Game Industry 181

13 Peer Workshops and Collaborative Writing 207

Illustrations

Acknowledgments

The authors would like to thank the following individuals for their input and spirit of collaboration in developing this book: Julian Whitney, Trent Hergenrader, Margot Douaihy, and Nat Mesnard. Thank you as well to Lucy Brown and all the editors and production staff at Bloomsbury Academic. Thanks to the Associated Writing Programs conference, the conference organizers, and all the past panelists, teachers, and students who continue to support this work. Lastly, thanks to the people behind the games and the stories that have moved and changed us.

1

Game Writing Is for Everyone

A Manifesto

Introduction

There's a reason you picked up this book.

Maybe a professor assigned it to you: "Required Texts," it says at the top of your syllabus. Or maybe you're interested in stories and want to expand the ways that you can tell one. Maybe you've played *Horizon: Zero Dawn,* or *Red Dead Redemption,* or *The Last of Us* and wondered, how did the writers come up with that? Maybe those AAA titles were a primer for something else, something less *Hollywood,* and you're just starting to find titles that challenge what you know about storytelling, games like *Papers, Please, Her Story,* or *Depression Quest,* interactive stories that have changed how you think about the world. Or maybe this book is just a suggestion initiated by some cookie stored on your computer: if you liked x, you may also like y—a journey that started from a subconscious association that's finally manifested itself in the book that you now hold in your hands.

Story Mode

The origin story behind *Story Mode* doesn't come from a singular moment. There are no dead parents or radioactive spiders. Instead, the genesis came from you, the students, the people most likely to be picking up this book. Several years ago, three students asked Eric, "Would you be willing to teach an independent study on writing for video games?" That moment led to others: a chance reading of "Braving the Controller," an essay that Juli published in the *Associated Writing Programs Chronicle* about her reasons for teaching writing for video games in the creative writing classroom. Her call to arms was the impetus for our first Associated Writing Programs panel on writing for video games, a panel we've repeated in various forms over the past several years, connecting with and combining with other educators who were doing what we were doing: teaching writing in what is becoming a dominant medium for telling stories in our culture today.

   Story Mode synthesizes our experiences, building on the work of other writers, gamers, and educators, to produce the first guide to writing for video games for the creative writing classroom. We've designed *Story Mode* around the core principle that games are for everyone and that understanding the creative process and conventions of good storytelling will help students write better games that are narrative-driven and inclusive, games that harness the power of language, of making choices, and of walking around in the skin of another. You can read *Story Mode* from the beginning, starting with this introduction, then reading our primer on genres and forms, moving to the chapters on creative craft and game design, straight through to praxis and tool suites in the game industry. Or you can flip ahead, each chapter like a side quest in an open-world game. Jump to "Setting and Worldbuilding" or "Designing Games for Empathy and Change." Whatever path you choose, we're happy to come along on the journey with you as guides, or peers, or sidekicks. We hope that it's as transformative for you as it has been for us.

Playing and Writing Games Is for Everyone

First and foremost, this *isn't* a manual for entering the games industry as a working professional. Some writers may use the exercises and techniques included within to build a portfolio in the goal of working on the next big-budget action thriller or dialogue-centric indie drama, but that isn't the primary intention of our book. That, of course, begs the question: What *is* the purpose of this book?

Story Mode: The Creative Writer's Guide to Narrative Video Game Design seeks to democratize video games as a medium—and that includes its history, play, and creation—using creative writing craft techniques that have been shaped by academic practice and theory for some seventy years. That means beginning on the ground floor, unpacking the dissonant history of game genres and narratives—in everything from cyberpunk dystopian shooters to text adventures simulating what depression feels like—before settling into the three main parts of our book: (1) creative writing craft and its complicated relationship to video games, (2) game design theory, and (3) overall praxis. A reader who comes to *Story Mode* with little to no knowledge of video games, creative writing, or even any specific technical know-how will, by book's end, possess a richer understanding of video games as a form and of how creative writing techniques honed by professionals can enrich nearly any electronic narrative. Readers will be able to write and publicly release their own games—complete with graphics and sound—and professors will possess a clear manual on installing everything from video game-centric workshops to small-scale group activities into their classrooms.

But the more important question is why?

What's the point of learning how to write a video game in the first place, especially if you aren't hoping to enter the games industry as a professional yourself?

Video games are not merely a means to zone out and pass the time. This reductive view has plagued the medium ever since *Pong* machines first became ubiquitous in sports bars and pinball arcades across the world. It's a retrograde stance that ignores, willfully or not, decades and decades worth of meaningful game stories written by both narrative teams employed by massive corporations and BIPOC creators working alone through the heat of the night. Video games are an incredible and unique medium for communicating ideas and stories because unlike static media like prose or film, the receiver of the art—the player, in the case of video games—is directly implicated in the narrative and constantly presented with choice after choice. This could be the most basic of decisions—should I move my tennis racket left or right to volley the ball—or it could be a choice that has massive effects on the overall narrative—how long should I delay bringing my dementia-suffering spouse to a full-time care facility? This ability to roleplay—not as a mere passenger as in the case of first-person prose or even in film but as a functional protagonist who makes impactful decisions—opens up a wide swath of brand-new creative possibilities for writers, and it also has a tremendous effect on audiences. Video games have the ability to put players into shoes and experiences that aren't their own, everything from LGBTQ+ teenagers navigating high school to indigenous warriors fighting against colonialism. That is an extremely powerful tool rhetorically, and it comes with an enormous amount of responsibility. In *Story Mode*, we attempt to unpack these under-explored ramifications and guide writers toward the creation of thoughtful games that provide meaningful experiences for the audiences you wish to speak to.

One of the first things we tell our students in video game writing classes is that their game narratives don't necessarily need to be expansive. Sure, you can aspire to create the next playable *Star Wars* or *Lord of the Rings* if you so choose—and we'll guide you toward crafting those types of games as well—but there also exists a rich lineage of deeply personal and smaller-scale narrative games, similar to the work of writers like Raymond Carver, Joan Didion, or ZZ Packer, that tell idiosyncratic stories that also provide news of the wider world. Games about raising children with cancer, navigating border crossings, or bristling against capitalism in a steel mill town all already exist. You don't need to work for a massive corporate studio to write and release games, and this is particularly important because of issues related to diversity, inclusion, and representation that are still very relevant to the industry. A video game can teach community members how to use a library or how to balance a budget. A video game can teach doctors how to interact with differently abled patients or can be used to draw people into DEI training or interoffice mentoring. *Story Mode* addresses all of these concerns while still examining approaches from more action-oriented games as well.

We've written this book because video games have the capacity to revolutionize both art and narrative consumption and creation. This is a fully interactive and immersive medium that is still in its infancy, far less than a century removed from its earliest conceptions. Already, we have landmark titles that showcase how far games have come as a storytelling medium—*Final Fantasy VI*, *Metal Gear Solid*, *Depression Quest*, *Cibele*, *Butterfly Soup*, the list goes on—and we believe that it will go even further as more and more voices create their own games, speaking to their chosen audiences and communities, conveying thoughts and ideas more varied and complicated than anything we've yet seen. For gaming as a medium to capitalize on its immense narrative possibilities, we need those games to exist.

We need you to write your games.

Our Approach

Part I: Creative Writing for Games

Writing for games shares DNA with other creative narrative forms. Part I of *Story Mode* identifies key components of good storytelling. Narrative games build upon these inherited elements. Games can also combine and blur genres due to their multimodal and interactive nature.

Chapters 3 and 4 focus on the building blocks of a narrative: character and conflict, along with story, plot, and interactivity. Chapter 3 first breaks down character into its most basic parts, showing how words shape more essential notions of character. Character becomes less fixed and more malleable depending on the words that the author uses to evoke them. The dynamic nature of character changes even more when introduced to conflict. As characters experience conflict, they further change and grow. Chapter 4 extends our discussion of conflict, looking at how story proceeds from the germination of character and conflict. Story and plot help us organize a character's journey by sequencing events to create an emotional impact. Games extend the emotional impact of narrative by introducing interactivity, by inviting the player to make choices and participate in a character's change.

Chapters 5 and 6 look at narrative tools such as setting, worldbuilding, and dialogue. Setting and worldbuilding are more than just the backdrop for a game but are tied to earlier concepts of character and story. They are integral to the gameplay. Like static narratives such as fiction, games use mood and tone to affect character and conflict. Interactivity shows how setting can actually activate the narrative, providing the character with interactive artifacts or tasks that inhabit the world. In the chapter on dialogue, we show how to write unique and interesting dialogue that is selective and true to character. Each one of these chapters provides analysis from mainstream and indie games to help you understand core concepts as well as exercises applying creative principles.

Part II: Game Design for Creative Writing

Part II of *Story Mode* dives into elements of game design and useful game studies concepts, with the overall goal of helping readers to understand the basics of how games work and how they are put together. It gives an overview of various video game genres that prioritize narrative before exploring game design concepts and basic terms and critical approaches that can be used to examine games and understand how their experiences are crafted. As creators, working in games can seem daunting. How is it possible to tell a narrative in a medium where it can be difficult to predict what players will do or how they will behave? Further, you'll learn things like strategies for crafting meaningful choice and player engagement. You'll see how even small game design changes can lead to drastic differences in play experience, and you'll learn how to adjust those elements to best suit the story you're trying to tell.

Part II also features a focus on representation, diversity, and inclusion. Games are frequently associated with problematic experiences, whether that be in-game representation issues such as oversexualized female characters or racial stereotypes, industry and workplace-related issues such as harassment or unequal treatment, or player community issues such as online harassment or hate-raiding. As creators, we must always be cognizant of the ways that our works engage with perspectives different from our own and be mindful of the responsibility we bear for making our works as inclusive and diverse as possible. You'll also learn useful tools for engaging with diversity and inclusion in the games you create as well as strategies for approaching these systemic and complex issues.

Finally, *Story Mode* emphasizes that not all games have the same goals and that the reasons why someone might create a game are varied and diverse. While some creators may want to write epic fantasy narratives, others may want to use games to communicate a personal experience or to help players understand an important problem facing our own world. The final chapter in this part offers tips for designing games engaged with real-world issues such as mental health, sustainability, poverty, or political struggle. Games can be powerful instruments for persuasion and communication, and Part II will help you use those tools to the fullest, regardless of your project's goals.

Part III: Practical Tools and Approaches

Part III of *Story Mode* takes all the other aspects of the book and shifts them directly into praxis. Here, we introduce a series of tool suites you can use to write and release games. This includes:

(1) Twine: perfect for building online text adventures reminiscent of choose-your-own-adventure novels.

(2) Audacity: an audio-editing platform used to craft everything from in-game audio diaries to soundscapes.

(3) Ren'Py: a visual novel editor.

(4) Bitsy: a design suite that allows you to create games with moving graphics and sound.

The chapters in Part III focus more on philosophy than direct technical instruction—although that too is provided—and we also include assignments from our own game classes that guide writers toward creating shapely fiction inside the confines of these design tools.

We then briefly explain what it means to work in the video game industry and how to get there. We consider the barriers and discrimination facing women and BIPOC creators, and we also unpack various industry jargon. What's the difference between an indie game and an AAA game or a game writer and a narrative designer? We explore various ways to construct a portfolio of work, including game jams and self-publishing. In addition to a graduate school, we also highlight the various job boards and networking opportunities where people most frequently land game jobs, whether that means working inside a massive corporate structure or within a small team of just a handful of people.

Finally, we conclude *Story Mode* with a history of the workshop in the creative writing space. What is a workshop, and why has it been justifiably criticized as a pedagogical tool that hampers the ability of BIPOC writers to both speak up and write to their chosen communities? We provide alternative models aiming to address this criticism while simultaneously exploring what it means to workshop a video game and why you should or should not attempt it. We also explore the benefits of rapid peer review on nascent video game projects along with the pros and cons of collaborative writing—the standard in the larger video games industry.

Conclusion and Call to Action

Regardless of your reasons for picking up *Story Mode*, we hope that you'll emerge from your journey through this book with a stronger sense of how video games work and how you might use them to tell important stories and communicate essential truths. Video games are a massive global industry, with billions of players worldwide. When we ask our students to talk about their favorite narratives, they frequently describe digital game experiences, listing the plot twists that surprised them, the characters that wound their way into their hearts, and the expansive, wondrous worlds they were reluctant to leave behind. Whether you've been playing games for your whole life or are coming to them for the first time, there's no disputing that these kinds of stories are memorable, evocative, and captivating, and offer the potential to convey experiences and ideas in unique ways that are difficult to achieve in other media forms.

To an outsider, navigating the game industry may seem daunting, but the good news is that you don't need a degree in computer programming or game studies to begin crafting your own digital narratives. All you need is time, determination, and the willingness to learn something new. Creating your own games and sharing them with others is a powerful and useful tool, one that can help you affect audiences in unprecedented new ways. Whether you're trying to teach players a useful real-life skill, to convince them to empathize with a particular perspective or experience, or to tell an unforgettable story that will stick with them forever, knowing how to write a game is an effective skill that will serve you well in a variety of creative situations. Online tools and communities, too, make designing, crafting, and sharing your game with others easier than ever before. As *Story Mode*'s authors, we've written this book because we love games. We've been captivated by digital stories and characters; we've experienced the thrill of a new level opening up before us; we've worked with online friends we've never met in person to finally defeat a seemingly unbeatable boss; we've all stayed up way too late. We believe in the unique power of games to tell stories in ways that are unprecedented, memorable, and compelling. Further, we believe that anyone (everyone!) can and should be able to use these strategies and skills in service of their own stories, no matter their background or abilities. In this book, we've done our best to give you the tools you need to craft your own digital games. We're so excited to see what you create.

2

Context and Foundations

A Definition

What is a video game? This deceptively simple question has tormented academics since video games first became an area of study. Maybe you find this silly. Isn't a video game whenever we push a button on a device and cause something to happen on a screen? The answer is sometimes, but not always. If a viewer is watching a sitcom on her television and uses the remote to flip to a drama, she's pushed a button on a device that caused something to happen on a screen. And yet no reasonable person would call this a video game. Conversely, if the same person is playing *Angry Birds* on their cell phone by swiping and pinching, she's clearly playing a video game even if there aren't any buttons involved.

Defining video games is a tricky business, but generally you recognize a video game when you see one. A robust example of definitions can be found in Katie Salen and Eric Zimmerman's *Rules of Play*, which breaks down numerous definitions of both types of games and types of play while comparing them to other definitions used throughout the field. For our purposes, however, let's define a video game as an object that meets three criteria:

(1) The player is presented with choices.
(2) There is a significant digital interface.
(3) There are rules.

This isn't a perfect definition, and already we can imagine examples that press upon this definition's outermost edges—we might recall the board game *The Omega Virus* with its five-button console or even 2009's *Noby Noby Boy*, a video game that's more of a play space where the player controls a worm simply wandering around. However, we believe our definition holds up even against the most extreme examples. Let's break them down one at a time.

(1) The player is presented with choices.

Choice is a slippery proposition in video games. You might immediately imagine some kind of grand narrative decision. For example, consider the choice in *Fallout 3* when the player must decide between detonating or disarming a nuclear weapon that is about to destroy a settlement. That's a fine way to think of choice, but narrative choices don't present themselves in every video game. When we're discussing choice, imagine everything from the aforementioned *Fallout 3* decision to something as simple as the opening screen of the original *Super Mario Bros.* Here, the player appears as Mario on a flat 2D plane. Waiting just ahead is a Goomba, the most basic foe in the Mario series. It shambles toward Mario, and the player is presented with a simple choice: jump and evade the Goomba or stomp and kill it. When you view choice through this light, it's obvious that video games present us with decisions over and over and over again. Are you moving a character? Are you guiding a cursor? Are you picking dialogue options? All of these are choices, and they're rarely used in static media like literature, film, or music.

(2) There is a significant digital interface.

The key word here is "significant." In our *Omega Virus* board game example, players move plastic pieces around a map surrounding a small device with five buttons and a speaker. Compare this to a PlayStation 5 or an Oculus or an Android phone or even the original GameBoy, and this is not a significant digital interface. Pinning down what is "significant" is difficult, but we typically mean multiple interface options—buttons, a diamond-shaped directional pad, perhaps even joysticks—and some type of screen—whether that be a TV, laptop screen, phone screen, dedicated screen, or even the two screens in front of your eyes in VR goggles. Without a significant digital interface, the game becomes something else—more akin to an analog game with minor electronic elements like *The Omega Virus*.

(3) There are rules.

This is perhaps the most self-explanatory guideline, but a game must have rules. Otherwise, it ceases to be a game and turns into something resembling static media. Rules, however, don't have to be defined by points or losing health or even causing a game over or fail state. Even in *Noby Noby Boy*, the player can't venture into the liminal space beyond the game's playfields. That's a rule.

A Brief History

Much like defining what a video game is, academics have struggled to pin down the exact starting point for video games as a medium. Some point to Ralph Baer, a German American engineer whose family fled Nazi Germany. While working for a defense contractor in 1966, Baer began constructing a wooden prototype that allowed children to play simple games on their TV screens. This technology was eventually licensed out and released in 1972 as the Magnavox Odyssey, a device that produced black-and-white graphics but no sound.

Others look to Steve Russell, a computer scientist at MIT who in 1962 developed *Spacewar!*—a simple interstellar cat and mouse game—only playable on the PDP-1, a then cutting-edge computer that was mostly only available at universities. Still others favor *Bertie the Brain*, an electronic game of Tic-Tac-Toe built for the 1950 Canadian National Exhibition by Josef Kates, an engineer who went on to develop the world's first automated traffic signal (Figure 1).

No matter which origin you choose, the conditions for the emergence of video games as a cultural phenomenon remain the same—the intersection of computers and screens in the university and military gave rise to engineers and inventors taking this technology and using it to write games. Nolan Bushnell, one of the students who first played *Spacewar!* in the 1960s, instantly recognized its mass appeal and began building a consumer version for bars. Reports vary on how successful Bushnell's *Spacewar!* knockoff really was upon its release in 1971, but the following year he attended a demonstration

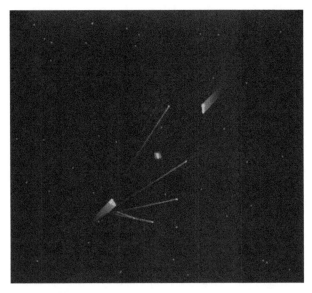

Figure 1 Is 1962's *Spacewar!* the first example of a video game?

of Ralph Baer's Magnavox Odyssey. There, the 29-year-old who had moved from Utah to the sunny shores of a then unnamed Silicon Valley first witnessed a black-and-white tennis game for two. He set his friend and collaborator, programmer Al Acorn, on creating an arcade copycat and later that year released *Pong* under the Atari banner. After that, everything changed.

Although there's disagreement about what the first video game is and who created it, there's little argument about the game and company that pushed video games into the mainstream—at least in North America. The arcade version of Atari's *Pong* was massively successful, generating 140–160 plays per day in some locations and selling units all over the world. By 1975, Atari began releasing home versions of *Pong*—dedicated machines that played only one game—and in 1977 released the Atari Video Computer System, sometimes called the VCS or 2600, which featured a cartridge slot where players could swap in new games as they became available at retail. Atari was by no means the first company to release a console with swappable cartridges. That technology was developed two years earlier by Jerry Lawson, a black engineer from Brooklyn raised by a longshoreman. Unfortunately for Lawson, his cartridge tech was utilized in the obscure Fairchild Channel F which was soon crushed by Atari fever, fueled even further after Time Warner purchased Atari and infused them with resources (Figure 2).

By the end of the 1970s, Atari had sold over one million copies of its Video Computer System and was buoyed into the next decade by adaptations

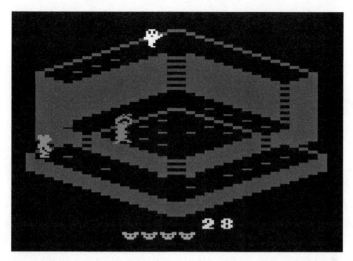

Figure 2 A 1984 home console adaptation of Atari's arcade classic *Crystal Castles* for the VCS.

of arcade hits retrofitted for the home market. That saturation, however, was Atari's undoing. By 1983, the video game market in America collapsed under the weight of shoddy, rushed-to-market games, culminating in the near apocryphal tale of thousands of copies of *E.T.*—coded in just four weeks by Howard Scott Warshaw who typically took six months to finish a single game—being discarded in a New Mexico desert. Although video games were initially dubbed a fad, Japanese company Nintendo soared to even greater heights with the release of its Nintendo Entertainment System console in 1985, which would go on to sell nearly sixty-two million copies worldwide.

With the establishment of Nintendo as a dominant cultural force—along with its Walt Disney-esque cast of recognizable characters including Mario, Luigi, Donkey Kong, Link, Zelda, and Kirby to name only a few—the once fringe video game cemented itself as a powerful medium that soon generated more money than the film and music industries combined. As sales increased, competitors to Nintendo emerged, most notably Sega and later Sony with its PlayStation consoles and Microsoft with the Xbox. We'd be remiss not to mention the massive computer and PC game boom that began in the 1970s and continues today. And don't forget the exploding mobile game market. In 2021, nearly half of all Americans played at least one video game on their phone. Video games are as ubiquitous now—or perhaps even more so—as watching TV was in the 1960s. A medium that once comprised one-screen games like *Pong* or *Space Invaders* now resemble epic films with orchestrated music, voice acting, and endless hours of gameplay. Like it or not, video games aren't in the process of becoming the dominant form of storytelling in the twenty-first century. They already are.

Getting Started

It's all well and good to define a video game and to understand a little bit about where this relatively young medium arises from. However, how do you play them? Where do you find them? How on earth do you even begin?

Today, video games are played across more and increasingly diverse platforms than ever before. As referenced earlier, one of the easiest places to begin exploring games is on a mobile phone or tablet. Here, typically in your phone's app store, you can access any number of games that rarely require controllers—another name for dedicated joysticks—or any input more complicated than swiping and clicking. For a simple place to begin, we recommend downloading *Florence* to your phones. This game follows 25-year-old Florence Yeoh and the rise and fall of her romantic relationship with a young musician named Krish. It's a tender, bittersweet game that often charms our students without overwhelming them with complicated inputs or gameplay concerns. *Monument Valley* is an equally compelling place to begin. In this mobile game, the narrative tracks Princess Ida as she explores an Escher-like environment in search of answers and forgiveness. This game is slightly more mechanically complicated—asking the player to solve a series of increasingly more difficult puzzles—but is still rather forgiving while providing a wonderful entry point for folks who may feel uncomfortable with video games as a medium.

For writers and instructors looking to venture beyond the mobile space, we'd next recommend the use of a computer. Whether it's a PC, Mac, desktop, or laptop doesn't really matter—although be aware that there are more game offerings for the PC. Either way, there are still copious examples of simple games you can play and discover on any of these platforms. We often recommend beginning with itch.io. We'll cover this more in-depth in the final chapter of *Story Mode*, but, in brief, itch.io is a website where you can play hundreds of games. Some are playable right in your internet browser, while others must be downloaded. Some are free, and some require payment or a donation. We'll begin by recommending a few games we cover elsewhere in more detail. The first is *Depression Quest*, a web browser text game simulating what it feels like to struggle with depression. Here, you play with your mouse, simply reading text and clicking choices—similar to most websites. After that, we'd recommend *Butterfly Soup*, a downloadable text game about a group of queer Asian American teens thriving in San Francisco. Neither of these games require a controller, and neither require quick reflexes.

If you're eager to try something more mechanically complicated on itch.io, we'd steer you toward *Celeste*, a two-dimensional platformer—similar in presentation to *Super Mario Bros.* or *Donkey Kong*—that features a story about depression, panic attacks, and overcoming self-doubt. Unlike our previous recommendations, *Celeste* prides itself on intense difficulty. Here, you might want to try using a controller or a joystick—you can find any number of them on sale for less than $20 by Googling "PC Controller" or "Mac Controller." Plug them into your computer via USB, follow the instructions, and you'll be good to go. Despite its difficulty, we've included *Celeste* on our list because of its commitment to accessibility. Here, you can tweak most of the game's difficulty options, tailor-making an experience that's satisfying for you.

Beyond itch.io, you might also experiment with Steam, a downloadable video game storefront that offers you just about every commercial release available for PC and Mac. While anyone can upload a game to itch.io, there's a strict approval process developers must go through to list their games on Steam. In terms of presentation, imagine iTunes, a digital storefront that also acts as a launchpad for your media. Just like itch.io, you can use Steam with or without a controller, and we'd recommend *Her Story*—a game that plays out more like a live-action movie—and *Stardew Valley*—a cozy game set on a farm with a surprisingly deep and relaxing narrative.

Beyond mobile phones and computers, you can also opt into dedicated video game consoles that often plug directly into your TV. This is the least cost-effective route, requiring players to shell out hundreds of dollars for a piece of equipment primarily for playing games. As of this writing, there are three significant console offerings: Sony's PlayStation 5, Microsoft's Xbox Series S and X, and the Nintendo Switch. Generally, most modern titles are playable on all three platforms. The PS5 and the Xbox Series X are the most powerful in terms of hardware, while the Series S and Switch are cheaper. The Switch also has the bonus of being portable, coming equipped with a handheld screen you can take on the go. These consoles can be purchased from most big box stores or online retailers, and they almost always come with at least one controller to get you started. Here, we'd recommend *Undertale* and *Firewatch*, both available on all the above platforms. For platform-specific offerings, we'd steer curious writers toward *Uncharted: Legacy of Thieves* on PlayStation 5, *Fire Emblem: Three Houses* on Nintendo Switch, and *Pentiment* on Xbox Series S and X.

Finally, there's the nascent world of virtual reality. Currently—and this space is exceptionally volatile as of this writing—there are two significant offerings: PC-centric VR headsets like Meta's Quest 2 and then Sony's PlayStation VR. Both are elaborate visors/controller sets, but the Quest 2 taps into Steam's vast database of games, while the PlayStation VR is less expensive but only works with a more limited ecosystem of offerings. At least for now, we only advise VR gaming for extremely dedicated video game players, but we'd recommend *Half-Life: Alyx* and *No Man's Sky VR* on the Steam options and *Blood and Truth* and *Batman: Arkham VR* on PlayStation.

Case Study: *Butterfly Soup* (2017)

When we ask our students for examples of excellent video game stories they love, they occasionally have difficulty generating answers. Some of them don't play a lot of games, and those who do sometimes enjoy sports games or first-person shooters online or with friends. Occasionally, we have students who eat, sleep, and breathe games, and they might throw out something like *NieR:Automata*, *Uncharted 2*, *Metal Gear Solid*, or *BioShock*. But even these examples typically revolve around violence. A lot of what we discuss during that first discussion session is stereotypical fare: guns, swords, and murder.

That is exactly what makes a game like *Butterfly Soup* so refreshing and such a joy to experience in the classroom. This visual novel written in Ren'Py by Brianna Lei revolves around four LGBTQ+ Asian American teens navigating the start of ninth grade. The game probes homophobia and even parental abuse, but the tone is effervescent and hilarious, and the dialogue pings across the screen as Akarsha references anime and makes dumb jokes and pulls gag after gag after gag, usually on the self-serious overachiever Noelle. The game primarily focuses on the burgeoning relationship between Diya—a quiet but gifted athlete—and Min-seo—the 5'1" firecracker who jokingly threatens to murder everyone in her way. Much of this romantic banter plays out on the baseball field or the locker room or the Indian buffet in their California bay-area neighborhood. Compared to characters in even some of the finest narrative games we've referenced earlier, these four teens feel like actual human beings. Their jokes and hardships resonate, and it's wonderful assigning *Butterfly Soup* and watching students learn that video games don't need to foreground violence. They too can explore the difficult and varied emotions that traditional media has been plunging for hundreds of years. It's also vitally important for people who feel like games aren't for them to play something subversively against the grain like *Butterfly Soup*. We've witnessed it in our own classes. Suddenly, students with little interest in games begin to feel like the medium holds space for them after all (Figure 3).

In terms of narrative structure, *Butterfly Soup* is laid out in an interesting way. The game takes about seven to eight hours to complete and there are no fail states. Players can reach the end no matter their skill level. The narrative crosscuts between the start of ninth grade and the summer around third grade when most of our characters first meet, and each of *Butterfly Soup*'s four chapters features a different girl as its protagonist. In most visual novels, you're tasked with making critical decisions that spiral into different narrative paths and endings. In *Butterfly Soup*, you typically choose between snarky wisecracks and not one of your decisions has any impact on the larger narrative whatsoever. Writer Brianna Lei understands the story she wishes to tell and the ways in which it pushes against typical video game narratives—especially those involving LGBTQ+ women. You can briefly deviate from the path—perhaps you'll prank Akarsha by

Figure 3 The humor of *Butterfly Soup* quickly gives way toward a deftly felt emotional core.

tricking her into triggering the school library's alarm—but the romance will play out the same way every time. This, too, is instructive. Too often, when we think of video game narrative, we think primarily of narrative choice and justifying whatever impulsive decision the player makes. But sometimes, especially when you're telling a personal or politically motivated story like *Butterfly Soup*, it makes sense to restrict that freedom and allow it to flourish in other ways that don't affect the main narrative.

While so many of the games praised for their narratives take cues from Hollywood blockbusters, *Butterfly Soup* feels thematically connected to writers like Misa Sugiura, Alison Bechdel, and Sally Rooney. It's the perfect entry for writers who want an example of realism in video games that eschews violence for humor and love, and it's a breath of fresh air for students who feel othered by so many mainstream games. We can't recommend it enough.

Overview

Defining a video game can be rather difficult, but, for our purposes, a video game is an artifact where the player is presented with choices, rules, and a significant digital interface. In this chapter, we've also briefly covered the vast history of video games, beginning with the medium's origins in *Spacewar!* and inventive creators like Jerry Lawson leading all the way to the present day with high-powered console systems ready for home use. If you're just beginning your journey in this often-overwhelming field, we recommend starting with itch.io and Steam before branching out to explore accessible console titles or even the more precarious and rapidly changing world of virtual reality.

Part I

Creative Writing for Games

3

Character and Conflict

It's November 23, 1982, your first day of work at a border checkpoint in authoritarian Arstotzka.

Weeks ago, the labor lottery landed you this job. The state relocated your family to a class-8 apartment building, one of those Soviet-like blocks of cement with identical-floor plan apartments stacked twenty stories high. But at least it was in the city. Better education, better health care, a place of possibility for your son and where your ailing mother-in-law and your uncle could grow old with dignity.

In the morning, you walk to your booth at the Grestin border checkpoint, its first day open. Lines of Arstotzka citizens, visitors, refugees, and immigrants snake into a concrete yard. A wall topped with barbed wire stretches to either side of you. Soldiers with machine guns patrol the walkway.

At first, your instructions are clear: only Arstotzka citizens. So many in the queue are hoping for more. They are foreign relatives, asylum seekers, people from poorer countries looking for work. You don't have time to hear their stories. You're paid by your processing volume, a euphemism to describe the very real people you face and reject.

Soon the rules change. First you let foreigners in. Then workers. Sometimes they need a special permit, sometimes not. The amount of information you're confirming slows you down. You interrogate people. Right gender? You realize this visa is expired. You need work authorization. Sorry, I don't accept bribes. Sometimes they tell you: *I haven't seen my son in years. I have a granddaughter I've never met. I need this job.* If you make a mistake, you're docked pay. You need money for food, heat, and for medicine when your feeble son inevitably falls ill.

It isn't getting any easier. When a suicide bomber—someone you let through—kills a group of Arstotzka soldiers, your job becomes more difficult. Your allegiance to Arstotzka starts to waver. There are consequences to being merciful. Soon, you have a choice: support a splinter group offering you safe passage out of Arstotzka in exchange for your assistance. Or stay the course, keeping the rules so your family can survive.

The PC gaming website Rock Paper Shotgun calls *Papers, Please* "engrossing, darkly ominous, and a fascinating exploration of morality versus progress." Developed by former Naughty Dog employee Lucas Pope, *Papers, Please* asks the player to step into the role of a lowly immigration officer in a repressive Eastern-bloc-like regime.

The environment and gameplay in *Papers, Please* shape your character. The first few days of the game you're able to dismiss the over-the-top patriotic slogans and settle into your immigration officer routine. Arstotzka iconography still bothers you, though, with its Nazi-like eagle and the stark red-and-black colors. But gradually elements of the regime start to infringe on your personal life. The money you make is never quite enough, and the first few run-throughs of the game leave your family cold and starving. When you're finally faced with a choice that could topple the government, you're willing to take it, even if it means that you and your family could face imprisonment or death. Still, you can choose to follow orders despite your difficulties. Your choices lead you to multiple endings. Your character depends on who you trust, how far you're willing to go, and what principles you apply to guide you through the game.

What makes a compelling character? If we ask our students about their favorite characters, they'll give us a variety of responses. They like Frodo Baggins, Katniss Everdeen, Sula Peace. And Batman and Solid Snake and Max Caulfield. They like Deadpool and Captain Marvel and Eleven from *Stranger Things*. When we ask why, it's harder for them to articulate. If it's a book, the character is inevitably tied to the story. If it's a film, sometimes it's the actor, their attitude and ability to deliver memorable lines. With all the examples a few commonalities emerge: it's what they *do*, their ideology. They're cool, these characters, which translates to identifiable and complex. They're also generally characters that our students want to be.

So how do you make characters like these? Before we get to fully-fleshed-out characters, let's think of character as image, character as figure, character as narrative line, character as word. In J. Hillis Miller's *Ariadne's Thread*, we're reminded of how language shapes what we read on the page or see on the screen. Miller says,

> Most English and American readers of novels, even those trained experts who are for one reason or another writing essays on a given novel, pass through the language of a novel as if it were transparent glass. They begin talking about the characters in the story as if they were real people, seen perhaps through that glass and perhaps distorted by it, but not created by language. (29)

Think back to most every English class you've ever had. How did you talk about character? When you finished the book and read those words into being, didn't you discuss the characters as though they had their own value systems and worldviews, as though at any moment you could have a conversation with them and know what they were thinking, or how they would act in a given situation? Because by now you've spent countless hours with these people, sharing their most intimate moments. You know what they think about their parents, how they never quite seem good enough, what it's like to live with the burden of a secret that you can't share without hurting those that you love. And you also learn the inverse: Frodo would never do anything like that! A moment when a character in a book or a film or a TV series starts to behave in ways that are incongruous with what you know about them. You identify with the character so much that you feel a connectedness, sometimes even an ownership. A lot of fan fiction is driven by readers who want to explore what a character they know could do. Reading through the language like it's transparent glass isn't a bad thing. As writers, this is kind of our goal. We want our readers to identify with our characters. It's when we want to know *how* to do it that looking at the glass becomes important. What makes a compelling character? Well, let's start by examining the words and digital mechanisms that we use to bring them to life.

Again, it's helpful to think of character not as a fixed, essential identity but as something that's malleable based on the words you choose. In *Craft in the Real World*, author Matthew Salesses writes:

> An exercise that has helped my students and myself with characterization, plot, world-building, and so forth, is to write a list of every decision a character makes in a story, in order, skipping nothing, not even what they choose to wear that day or negative choices (things they choose *not* to do). (69)

To further explore character, try a word-based exercise to break down the connection between a word and what it evokes. Ferdinand de Saussure, the structuralist literary theorist, called this sign and signified. Most of the time, this relationship between the word and its meaning runs under the radar. We just assume that the word "table" is a table, that our visual implied meaning is similar enough to others around us to make sense of it. Rarely do writers need to question this relationship but instead use it to their artistic advantage, choosing words that in their descriptive precision provide the reader with pleasure and surprise.

An early exercise that we will use to get students thinking more broadly about character is called the Character Lottery, a Mad-Libs-style character creator exercise. You can do this individually or in a writing room, generated by either an instructor or members of a group. We prepare for the exercise by collecting several characters on slips of paper with a name, age, and occupation. After, another group of abstract characteristics. Then a group of desires or goals. We will pass around each of these in hats or some other selection device (we often hold them face-down like little playing cards). You pull these at random. You could get Gary Brickman, 33, engineer for CAT, or Charlene Thompson, 25, aerobics instructor, or Marla Mathews, 50, organic farmer. After that, you pick an abstract characteristic. You could choose paranoid, homesick, or infatuated. Then a desire: the character wants to travel to Europe, be on a reality TV show, or fix everyone's problems. The exercise forces you to allow changes to character based on random words and associations. These tend to disrupt stereotypical or stock characters much more than if your teacher had asked you to write simply about an aerobics instructor. You have to allow the character to adopt characteristics that you wouldn't come up with on your own. Now you have to put the character in conflict showing them trying to achieve their desire. The result will often be characters who are more complex and interesting, who will push the narrative forward in ways that you couldn't have anticipated.

But some concept of character unity is necessary. This exercise only works if the elements that you are provided with are specific and interesting enough to make it worthwhile. Often when we do this exercise, students will pick a characteristic and they'll look at it, perplexed. How do I write about a racist toddler who wants to buy a pickup truck? We will allow students to repick if the emerging character is so disrupted that an image or story refuses to emerge.

Depending on the class and your professor's teaching style, you may complete this on your own outside of class or draft it in class and then revise it later. Inevitably, you'll end up with a character that surprises you and that is different from what you may have come up with had you simply been asked to write an interesting character. It also moves the discussion away from essentialist notions of character, that is, who a character *is*, and more toward the language that we use to evoke them. You might have to do a little research. If your character is a urologist and you have no idea what a urologist is and what they do for work, it will affect your ability to write the character in an interesting way. It also requires you to look more closely at desire. Characters are more interesting if they have goals, if they risk failure, and if they're not one-dimensional. This exercise will give you experience writing desires or goals that aren't so enormous as to become meaningless. The veneer of language requires a lot of tempering if you want it to appear like transparent glass.

This in-class gamification of character creation can also be helpful for getting you to think about characters in video games. For gamers, the concept of character as something fixed and immutable is often less a problem. They expect their character to change, altered by either their choices or their environment. Avatars are powerful expressions of character, allowing gamers to choose someone who may look and behave like themselves or to wander around in the skin of somebody completely different. Games often relinquish control of a character to the player. Instead of randomly picking characteristics out of a hat, they are the puppet masters, thinking about what options for customization could result in compelling gameplay. In these situations, game developers are thinking about what will allow the player to have the most satisfying game experience with the characteristics that they have selected. This exercise then can also get you to think about the kinds of characteristics you want to have available to a player. Imagine instead of being tasked with having to create a character from a random group of words and desires to being the person to choose the characteristics to begin with. Your job is to choose specific and unique characteristics so that the player will have interesting customizable choices available to them. The more interesting these characteristics, the more interesting their character will be.

Sometimes characters in video games are entirely defined by player choice and not by a writer. Sandbox games, for example, often provide very few defining characteristics, instead allowing the player to develop a character themselves. In some games, that's literally the point. Who will your character be as you choose what to do? In these situations, a writer needs to be less concerned with supplying the player with fixed characteristics and more with how to provide meaningful choices or compelling tasks that will allow the player to develop the playable character.

Complicating Character

But in situations where the writer isn't relinquishing power to the player, character needs to be nuanced and complex. One common problem with student character creation is the "everyone" character. The student's intentions are good: they want to be inclusive. This character could be anyone! But what happens, almost inevitably, is that the character ends up bland with a lack of personality or discernable goals. So we push. What does an "anyone" character look like? Act like? What beliefs or values do they have? Who are their friends? If the student still insists on generalities, the character will be stereotypical: a stereotypical waitress, a stereotypical schoolteacher, a stereotypical firefighter. These stereotypes or caricatures become even more problematic when you figure in gender and race. When we generalize about characteristics, what we're really doing is making a judgment about a group of people. And that can be dangerous. For some folks, it's a definition of bigotry.

To avoid generalizations, instead focus on the particulars. The Character Lottery exercise can help with this but it's still very bare bones: you only have a name, age, and occupation. Say you have a 23-year-old trapeze artist named Lily. What are some of her likes/dislikes? Maybe she dislikes mushrooms. She's a failed gymnast and that clouds her success in Vegas. She sometimes gets debilitating stage fright even though she's done her routine thousands of times. Sometimes when she's up on the high bar, the tips of her fingers go numb.

In games, myriad possibilities exist for fleshing out character particulars. Opening up the world allows players to interact with their environment to learn more about them. Writer Nina Freeman's 2015 game *Cibele* does this beautifully. We open on a specific date (February 18, 2009) in a young woman named Nina's dorm room. In live-action cutscenes, we glimpse Nina's pink hair, her manga collection, anime posters celebrating *Macross*, *Hello Kitty* stuffed animals, and eventually are given access to her PC Desktop complete with a *Sailor Moon*-esque wallpaper and folders earmarked for selfies and poems. As we scan the pictures and read Nina's poems, diary entries, and chat logs, we learn more about her, including particulars: she fetishizes Japanese culture, and she's inexperienced in manners of sex and envious of her more knowledgeable peers that the majority of her social life comes in the form of an online video game called *Valtameri*. Eventually, the player logs onto *Valtameri* and meets Blake, Nina's long-distance love interest.

The focus on particulars in *Cibele* eventually allows the player to imagine the totality of Nina's life. Nina's sexual curiosity, how she avoids in-person social gatherings, how she eventually invites Blake to fly across the country to meet her. Interestingly enough, the roles of Nina and Blake are played in the live-action scenes by the game's writer Nina Freeman and its programmer Emmett Butler. Nina's unique character traits, all set in a particular time and place (A New York City dormitory in 2009), allow the developer to lead the player to conclusions, to gradually uncover Nina's first brush with sex, culminating in her rejection by Blake. The particulars create a collage of an interesting character which in turn influences the story.

Character Desires

The other way to complicate character is to focus on a character's desires. Desires are even more closely tied to story since these are the character elements that will influence a character's actions. If a character's desires are complex and interesting enough, we're going to want to follow or play them in pursuit of those desires. Make the desires too trite, too easy, or too abstract and you're likely going to lose your reader or player playing the game.

If a character's desire is persuasive, it develops the character's *ethos*, a term Aristotle used to describe credence through authority. Ethos doesn't necessarily imply a certain morality, but that the character's desires match what we know about them in a believable way. Make the character too good or too evil, and we instantly begin to question their motivations. Also be wary of developing characters whose goals are too narrow or perfectly articulated. People are complex and often subconscious desires complicate conscious goals. If the character seems too one-sided or motivated by a single emotion—greed, revenge, romantic love—the player will either tune out or have a less satisfying gaming experience. Complex characters require complex desires.

Take, for example, Jodie Holmes, the protagonist in Quantic Dream's *Beyond: Two Souls*. From the beginning, Jodie is conflicted, psychically tied to a mysterious entity that sometimes wreaks havoc in her world. Her complicated relationship with the entity—Is she possessed? Is it some sort of telekinesis? How does she control it? Can she? And will she ever be able to live a normal life?—guides the narrative and the rest of the gameplay. These subconscious desires affect her conscious goals—they determine how she and the player react to the formative events of her life: leaving her foster family, growing close to a father-figure-like doctor, and eventual recruitment by the CIA. The seeds of her character complicate her desires and how the player navigates relationships. The player becomes mistrustful of others' motivations. Do they really care about Jodie or are they only using her to pursue their own self-interests? Her complex desires lead the player through difficult choices, which result in multiple, equally satisfying endings.

Sometimes desire become more complicated during the game. Aloy at the beginning of *Horizon Zero Dawn* wants simply to find out who she is and to become part of the Nora tribe. When she learns more about herself, those desires shift: now she wants to use her identity and the skills that she has acquired to help her people, to save her world from destruction. The stakes grow and develop during gameplay. What would've been a game about a woman centered on more selfish concerns—finding out who she is and gaining acceptance in her community—becomes a game about a woman selflessly working for the good of her people. Her desires have shifted, causing her character to grow.

Characters Must Change

As characters pursue their desires, does that journey (physical/spiritual/emotional) prompt change? Are the characters any different from when they started? Does their change seem real? Do you believe it?

One of the real temptations when writing characters is to have them completely overcome some flaw or problem. In *Making Shapely Fiction*, Jerome Stern identifies several "don'ts" of story writing that the beginning writer should try to avoid. One is the "Zero to 100 story," where a character overcomes some problem so completely that the reader questions its sincerity: the abusive spouse who suddenly loves her partner or the selfish miser who becomes a monk and gives his money to the poor. These narratives of complete 180-degree change are especially difficult to accomplish in a medium like the short story where narrative compression makes real change difficult. But the same concept can apply to longer forms. Stern explains,

> Massive character change is a staple of commercial entertainment. Half-hour situation comedies or one-hour mystery shows rely over and over again on a formula in which various family members finally realize they love each other or have behaved badly, but now everything is all right. These endings are emotionally attractive but, deep down, we know they just aren't true. (75-6)

Having a character undergo real, sustained change can be a difficult thing to do well. When a video game makes change too easy, the gamer feels cheated. The more difficult change is, the more the gamer is likely to feel transformed. Sometimes withholding change can also have a tangible impact, since gamers have become so conditioned to expecting it. Take, for example, the indie Twine game, *Depression Quest*. Just the title of the game implies a dynamic experience. What is a quest but a journey of change? But the player is quickly disabused of this notion. *Depression Quest* treats depression seriously, resulting in character choices that are often crossed out. The player *can't* choose to suddenly hunker down and get to work. The playable character may want to, but the mental illness of depression impedes their ability to do so. If the writers were to allow players to choose to get out of bed, get the promotion, work and smile till all was well with the world, it would feel disingenuous and fake. The player also wouldn't learn or empathize with how debilitating depression can be.

Shaping Character through Choice

When Eric was a graduate student at Ohio University in the early 2000s, he taught an upper-level fiction writing class that was based on the concept of choice in fiction. He wanted to deconstruct the notion of a unified character, to show how character could be multiple, complex, malleable, based on the words and choices that we used to evoke her. With the help of an open share web-based program, he developed "Storeeze," a choose-your-own-adventure-like online story space that no so cleverly combined "Story" with the spelling of his last name. Every week, a student would add another "Chapter" or series of choices or exploration of a thread to the story. Students could then choose one to four different choices, and the web program would automatically populate another web page that the next student could pick up and add to. When Eric designed the course, he imagined that their discussions would focus on how/why students made certain choices. Why did we pursue one path and neglect others? How did the character develop? By realizing how choice could show that character was multiple, students would develop empathy and understanding for others unlike themselves.

But that wasn't what happened.

Instead, students were intoxicated by choice, often making decisions that thrust the narrative into different directions. Some students would deliberately make choices to counteract others that preceded them. Sometimes the exercise devolved into a joke. Students thought very little about who their characters were, why they would make particular choices, and how their choices affected their notion of the character down the line.

In writing for video games, you need to approach the concept of choice in character very carefully. Games expand notions of character through choice. They play with, expose, and subvert traditional narrative forms, resulting in space for the player/reader to identify with multiple subjectivities. But there are potential minefields if choice becomes too easy or dissociated from character.

This is one of the most central challenges of writing narrative video games. How do you provide choices that fit with the player's developing sense of the character? Branching narratives in games can be massive, requiring a structure with story beats that are decided beforehand and workshopped in a collaborative writing environment. But the character still needs to make choices that both surprise and interest the writer and player. A playthrough of a full-length branching narrative, games like the *Life Is Strange* series, *Tell Me Why*, or *Mass Effect*, can take dozens of hours. Satisfying games with characters who change based on player choice can result in multiple playthroughs as players try other narrative lines to see what would've happened had they chosen differently. This is one of the strengths of the video game medium. The player is immersed in a world with complex choices. What would happen if I did this, or this, or this? Who would I be?

So, what does this mean in real-world application?

Consider the following section from a student Twine game. Twine, much like Eric's "Storeeze" game, allows you to create text-based branching narratives with ease. Here, the student has written a scene with a character suffering from addiction:

My limbs were numb and I could not move, even if I wanted to. My arms were heavy as if God laid stones on them. I had little control over my body. My physical body was trapped by a set of invisible chains but my mind was left out to roam freely. I rubbed my chest as it filled with warmth and I slowly cuddled into a ball on the beanbag. Pink Floyd was playing in the background, as my body fell comfortably numb. I wanted that moment to last forever because for once in my life, I did not worry about anything at all.

"Round two?" Colin asked.

I nodded my head with all the physical strength I had left. I glanced at a pair of shiny spoons on the ash-filled coffee table. I heard the click from Colin's lighter and the rattle of a plastic bag. The flame warmed the air and produced a shadow across the coffee table, which over time turned the spoon into a light gloomy red. The spoon became flaming hot and its contents from the bag boiled into a syrupy goo. The syrup popped, boiled, and turned into a sticky warm wax.

I closed my eyes, laid out my right arm and I felt the cold alcohol swab on my arm. Colin grabbed a thread, tied up my arm and sucked up the goo in a hopefully sterile needle. I felt the pin prick on my arm. Colin pushed the needle and pulled the thread.

The student then ends the section with two options: A. Let Colin do his job. B. Stop and get out of there.

So, what does X do? The student has set the scene, being careful to show familiarity with the environment. Colin says "Round two," implying that X has already had round one so another should be on its way. From what we know about the character, choosing A fits. X is an addict and by choosing A, we're confirming what everything up to that point has shown us. We want the character to choose B. But X is here for a fix and to choose B would be unnatural, uncharacteristic, unlike someone who has deliberately come to this place for exactly this purpose. So, what do we need for X to choose B?

This is where all discussions of character essentialism start to take the stage in a creative writing workshop. Character X wouldn't choose B because character X isn't like that. Where do we know this? From memoirs about addiction, from reading lots and lots of books, from personal experience maybe. We know that for character X to choose B, he would need therapy, an intervention, something that would affect his ethos enough to allow him that choice.

Pivotal choices like these require writers to rethink how they've planned narrative choice over the course of a game. We'll talk more about structure in our next chapter on plot but for now consider how much character and plot are intertwined. In order for a journey of change to be believable, characters need choices that deepen our understanding of them but that still may surprise us. This can become difficult if the game is overdetermined or too focused on one particular end result. The triumph of overcoming obstacles can sometimes feel cheap, especially with endings that are overly sentimental or that too easily resolve complex problems. Too many games unfortunately struggle with this. When dramatic change is too calculated, it can feel forced.

Sometimes, this can feel like a paradox: on the one hand, you've written a character with a planned character arc. We know how and when they need to change. And yet we also need the character to surprise us, to do things that are interesting and unique. Often our own biases can get in the way of figuring out what our characters can or can't do, limiting their potential. Would a character behave a certain way? Who are they really? Racism, privilege, and audience often factor into issues relating to surprise and believability. Matthew Salesses writes:

> The question of who believes something happened or not comes up a lot in life. It usually has to do with privilege. Again, microaggressions can serve as a useful example. It's not unusual, in my experience, that to [write] about a microaggression to a white audience is to have it or its racism called into question. This happens in workshop as it happens on the internet. I have been in multiple workshops where white students have basically said either "No one is that bad" or "That isn't so bad." (82)

So how does one deal with any of these issues, particularly in games, where players may have a lot of influence on how a character or story develops? How do you write a character with a planned character arc or even multiple character arcs of change and yet provide choices that will be both surprising and interesting? Here are a few suggestions:

(1) Don't force either/or choices that are unnecessarily polarized as purely good or purely evil.

Often in a story-driven game, we'll come across moments like this where the choice can be too black and white. The student has imagined the scene and described the characters but when it comes to making a choice at the end, the result is too choose-your-adventure-like, with the two choices being opposites of each other. The student thinks: Ok, now the character has to open the door. They want the story to move forward, and opening the door is the only way to do that. So, what would be an alternative to opening the door? Ah yes, closing the door. Playing a game like this feels like a life full of either/or choice fallacies. We either get one or its opposite when we should have choices that expand the world and its characters rather than limit them. Another problem with choices like this is that often only one of the two choices is plausible or builds from the scene that precedes it.

(2)   Use choice to further explore/develop character/the world.

This is related to the first one. Games like Dontnod's *Life Is Strange* (2015), a character-driven game about a photography student who can affect the flow of time, expands the world through choice. Look at a tote bag and discover its hidden significance to its owner. Look at a photograph to find out more about the main character's family. These kinds of choices are easy to make and they can loop back to one of the original story lines rather than produce new lines of action. Players may still end up making a choice that feels premature but the more that they're able to explore the character beforehand, the more prepared they'll feel to make them. In *Life Is Strange*, the more that we learn about the playable character Max, the better we're equipped to make decisions later on in the game. The mechanic of being able to rewind time also allows us to expand her capabilities, to help her grow in confidence and self-awareness. Some players may still avoid side exploration but providing this kind of horizontal character development will help the player deepen her experience and increase playability. What the writer needs to consider then is how to make choices that expand our notion of the character. These can be options at the end of a scene that the player can elect to choose or neglect depending on their playing style or investment in the world around them. Or they can be choices as the character interacts with their environment. Writers may use these choices to embed information that can benefit gameplay later on. The writer must then decide whether or not these world-expanding diversions are worth it. Too much side exploration doesn't move the character forward.

(3)   Know when a character arc has run its course.

One of the most significant difficulties in branching narratives is knowing when a particular line of character development is over. You've started your game with branches that may expand exponentially, sometimes leaving truncated narrative arcs. The characters in these arcs are similarly stunted. In these situations, ask yourself: Is it even worth developing the character along those lines? Is the choice that I provided offering the character any opportunity for growth? Or does it alter the character so much as to become meaningless? Not all character arcs need to have the promise of profound change, but they should be dynamic. If they're not, cut them. You can focus your energy on other narrative arcs. Which developments are most satisfying? Which feel like treading water? What provokes the character to do surprising or interesting things, prompting new growth?

(4)   Avoid character arcs that always result in death.

Resist the temptation when cutting truncated narrative lines to artificially end a certain narrative arc by killing the protagonist. This might seem counterintuitive in a genre dominated by the mechanic of shooting and killing things that move (at least in the AAA games). But for character development and for a more satisfying gaming experience simply killing off the protagonist for the hell of it can feel arbitrary, like throwing in a bunch of deus ex machinasto close off loose ends. Deus ex machina literally means "god in the machine" and describes early Greek plays that would lower gods onto stage using pulleys or some other mechanical device to conclude the play. Aristotle first used this term to describe artificially ending a play with any extraordinary event that's imposed on the narrative. Character death can often feel like a deus ex machina and shows a lack of imagination on the part of the writer. Don't know what to do when a character arc isn't developing? Have her step on a landmine! Boom, finished. The proliferation of endings like this in a character-driven story weakens the player's investment in the character. If the character's life doesn't mean much to the writer, it's unlikely that it will mean much to the player.

Point of View

In fiction, point of view (POV) is usually conveyed through the pronouns I, you, he, she, or they. Each has their distinct advantages or disadvantages. Writers like first person for its immediacy, for the ability to capture voice or character idiosyncrasies. First-person narratives often employ unreliable narrators since the narrator is biased and may not be very self-aware. Third person is the most conventional and flexible, allowing the writer to be both close to and far from the main character, depending on what they're trying to accomplish in a given scene. Second person is more experimental, often employing contexts that rely on second person such as an employee orientation or as a more self-reflexive stand-in for a first or third-person narrator.

In video games, first, second, and third person mostly describe a visual relationship with the player. In first-person games, the player sees what the character sees. Move the controller up and the screen moves up, like you're looking through a character's eyes to experience a virtual gaming world. In third person, the playable character is visible on screen. As you move the controller, the character responds like a puppet on a string. Second person is generally limited to text-based games where the "you" stands for the player, although just a few experimental visual examples exist.

As a window into character development, the choice of POV is an important one. First person became popular with the first 3D video games in the 1970s and 1980s, which were often maze-like affairs with the primary mechanic of shooting obstacles or enemies. The very first was *Maze War* in 1973, installed on computers at the NASA Ames Research Center. Like most early video games, mechanics dominated gameplay. First person was a way to introduce puzzles and levels by forcing the gameplay through a basic maze-like environment. First person did little to address *who* was manipulating the screen or to ask about what the relationship was between the player and the environment. It merely served as an interface. Later iterations of 3D first-person games such as *Doom* in 1993 introduced character elements with improved graphics and gameplay. These were often still games that relied on a shooting mechanic, now called first-person shooters (FPS). The reason for this is tied to the advantages of the first-person character. Like first-person fiction, first-person video games can create a sense of immediacy. FPS games capitalize on this advantage. They rely on it to build tension, to create the feeling of danger as enemies arrive in the FPS player's frame of view. Threats seem more imminent in first person. A first-person POV also mimics what a shooter may see were they looking through a scope to take aim at a target.

Many contemporary games have used the immediacy of first person to further develop a character. Games where the player's relationship to the gaming environment is in question or where the identity of the player is at first mysterious or ambiguous also may rely on first person. *Firewatch*, for example, uses first person to explore the Shoshone National Forest through the eyes of Henry, a novice fire lookout whose only contact with the outside world is through a walkie-talkie. Alone in a new environment, the first-person player navigates through Henry's world, encountering surprises or challenges that heighten Henry's reaction to them. When his supervisor Delilah, via walkie-talkie, tells him to investigate some illegal fireworks, Henry must confront some bikini-clad partiers who accuse him of leering. Later, he finds his watch tower ransacked. He sees shadowy figures and the girls he confronted are reported missing. The first-person POV contributes to a growing anxiety about Henry's presence; every choice he makes seems to heighten the tension around him. Because we're in Henry's POV, this has an increasingly destabilizing effect on the player. We're concerned because we empathize with Henry. We're just as trapped in the world of *Firewatch* as a marine fighting demons on the moon (*Doom*).

Third-person games, like third-person fiction, have become the most conventional point of view for story-driven games. Third person is especially helpful for developing characters with linear story arcs, one of the reasons it's so popular with AAA franchises where plot and narrative take center stage. It's the de facto point of view of choice for writers who want to exercise more control over the character, who have a definite story that they want to tell.

Third-person games have a long history. Early role-playing games (RPGs) or arcade games were 2D third-person games where you'd move around a pixelated character through mazes and dungeons. Whether you were Pac Man chomping dots of light or Mario pouncing on mushrooms, you played in third person, controlling the character you saw represented on the screen. As technology became more sophisticated, these third-person characters often became more realistic, moving from a 2D plane to a 3D immersive experience. Controlling a character now is a little like being a character in a film.

One of the advantages of third person is the ease of interspersing cutscenes with interactive gameplay. Cutscenes allow the writer to imagine pivotal moments with scripted dialogue and description. Third person plays very cinematically. The character usually appears on the screen in a full shot (head to toe) ready to be manipulated by the player holding the controller. As graphics, voice acting, and narrative have improved in these 3D-rendered games, comparisons to film have increased. We identify with the protagonist moving on the screen overcoming challenges the same way that we do while watching a movie. With cutscenes, we're content to sit and watch the drama unfold. When control returns to the player, our experience moving the character in the interactive environment increases our identification with whatever narrative challenges are thrown their way. Third-person games are more than just playable movies, although they share more DNA with film than any other type of narrative game.

Some games allow the player to toggle between points of view to suit their playing style. This has become more common with technological advances and with an industry that increasingly invites customization. Some players simply *like* the feel of immediacy that they get with a first-person perspective, while others may prefer to see their character move. Yet others may simply want to toggle between the two for certain segments of the game.

Meet the NPC: Other Characters in Video Games

Most of our discussion about character thus far has addressed the protagonist or playable character in a game. While this remains the most important character and the one whose experience is most tied to the story, non-playable characters or NPCs are equally important. They can provide background and texture to a game. They are the lovers, mentors, companions, or enemies that help the playable character realize or inhibit their goals. Interacting with NPCs can be distant and impersonal or intimate and life-changing.

With NPCs, the same rules of character creation apply. As you populate your story world with bakers, mechanics, shop owners, and police officers, it can be tempting to resort to the economy of stereotype. A game where nurses are too often women, drug dealers too often people of color, or love interests too often white with idealized bodies hasn't done enough to interrogate the socially constructed ways that we've become accustomed to seeing people. On collaborative projects, this is where having a diverse group of writers is so essential. It's less likely that you'll produce dangerous stereotypes. Workshopping also can help detect stereotypes. But not all the time. Some stereotypes are so pervasive that they can be hard to eliminate.

One tool to help avoid stereotypes are character sheets. Writing a detailed character sheet forces the writer to think more closely about the character, their motivations and desires, so that they become more fleshed out and round. Character sheets are also helpful for larger games with several writers. They provide a reference so that individuals working on different branches of a game are better able to write an NPC, especially as she relates to the playable character. A good character sheet should be nuanced and varied, revealing complex characteristics that expand our sense of the character and make her more interesting. The Character Lottery exercise can be a helpful starting point for a character sheet to randomize some of the characteristics so you're not forcing the character into a preconceived package about who they should be.

Character sheets can be simple or complex, depending on the game and the character's importance. Games generally will have several levels of characters, from simple, stock characters who may not do much to dynamic NPCs who are closely tied to the playable character with their own character arcs and goals that the playable character may influence through their choices. Even a relatively minor NPC can be dynamic as she interacts with the playable character. The more detail you provide in a character sheet will help identify potential avenues of tension that may arise in the interactive space of a narrative video game. A simple character sheet for a stock character may look something like this:

Stock character sheet	
Name, age	
Occupation	
Appearance	
Characteristics	
Goals	
Relationship to other characters	

Each category can be fairly basic, enough to give us an idea of who the character is and how she fits among the other characters or to the rest of the game. You can then adapt this simple template for more complex characters, adding specifics for each category to better flesh out the character. Under each category, you may have subheadings or prompts to standardize the kinds of things that will be important to remember or that have relevance to the game. Under "Characteristics," for example, you may have positive and negative personality traits, meaningful life events, or fears, joys, strengths, or weaknesses. Prompts could help us know how a character would behave in certain situations: the kinds of people that the character may be drawn to, or pet peeves or triggers for various emotions. Work up character sheets that have a fair amount of standardization across them so you can similarly flesh out your main characters and rely on them for later reference while developing gameplay.

Overview

In this chapter, we've shown how to develop compelling characters and how language, environment, desire, and choice can shape character in games. As characters pursue their desires, they encounter conflict, forcing the character to act. Aristotle wrote that plot is character revealed by action. In the next chapter, we'll explore the elements of plot in more detail, looking at common structures to chart your characters' actions. We'll also ask the question: Do narrative games need to have a plot?

4

Story, Plot, and Interactivity

Stars. Lots of them. Pixels of light grow brighter as you move through them. Fog-like nebulae alter the color from lavender to yellow to green. You pan left and right, searching the sky. Occasionally a typed name will appear under a star, marking it with a circle like an entomologist's pin. Finally, a destination. The screen goes white.

Initializing.

The white sky gradually fills with a rocky purple and brown landscape. Coral-like geodesic domes and conical trees with palm fronds clump together like mini oases on the surface of the harsh planet. Where are you? Umbrella-sized mushrooms provide a kind of ground cover in spots. It's then that you see your ship, smoke leaking from a breach in the hull. And something else: in the distance, movement, animal-like. They're small but as you get closer, you see the four-legged creatures with humps on their backs, like a cross between a turtle and a goat. Wherever you are, it's decidedly alien, like no place you've ever been before.

When *No Man's Sky* first launched in 2016, it was one of the most anticipated games of all time. Conceptually, the game sounded fantastic:

explore a universe that continually expands through computer-aided procedural generation, ensuring limitless new worlds with unique flora and fauna. This was the kind of game that epitomized the *Star Trek* ethos, to boldly go where no one has gone before. This would be a game of hours of unpredictable nuanced gameplay, the kind of game where you could truly lose yourself in another world.

But that wasn't what happened.

Instead, the game risked becoming a major flop. Players complained about the long hours traveling from one place to another and the lack of any discernible goals. Exploring turned out to be a very solitary affair. The algorithms for the procedural generation created world after world that differed only in the changing colors or flora or the new non-sentient species that inhabited them. The countless hours mining for resources and keeping your space suit and ship functional were the only moderate challenges in a universe that seemed mute to your presence. The maw of eternity opened up, filled with seemingly purposeless tasks and no one to talk to.

Financially at first, the game succeeded. The game was different from anything else before it, and the industry rewarded the concept handsomely. The 2016 launch was filled with unprecedented hype. This could be the game to change gaming forever, a self-generating sandbox that would literally create new worlds. But as more and more players traveled from star system to star system, counting down the minutes, they started to question the sixty bucks that they plopped down for the experience.

Why did *No Man's Sky* initially fail? Besides the botched rollout and the promises about the gaming experience that weren't fulfilled, the game lacked one crucial element.

A plot.

What exactly is plot? In the screenwriting book *Story*, Robert McKee describes a difference between Hollywood and European films:

> A typical European film opens with golden, sunlit clouds. Cut to even more splendid, bouffant clouds. Cut again to yet more magnificent, rubescent clouds. A Hollywood film opens with golden, billowing clouds. In the second shot a 747 jumbo jet comes out of the clouds. In the third, it explodes. (204)

The joke here is that European films often focus wholly on aesthetics, what the frame looks like and the image/characters/landscapes within that frame. The American film, however, is *all* causality. Something happens which causes something else to happen, often to the point of gratuitousness. Everything leads to the next event with time compressed to produce the most powerful impact, sometimes in a ludicrously short period of time. The differences between European and American film traditions are much more nuanced than what McKee presents here, but his oversimplification emphasizes one of the key components of storytelling: a plot.

First, a quick definition. A story is an account or narrative of events. Plot is the deliberate sequencing of those events to produce an impact. In other words, story is *what* happens and plot is *how* it happens. You've probably seen various models for this, sometimes with a visual aid like the reverse-check mark included in Figure 4. This shape is often attributed to nineteenth-century German playwright Gustav Freytag, and it's called "Freytag's pyramid" or "Freytag's triangle." The shape has several specific components: an inciting incident, a rise in action, a climax, and then a denouement. You can superimpose this rudimentary plot on the vast majority of stories we encounter in the world, from full-length novels to screenplays, to even your average video game. And there's a reason for this. Storytelling is a deeply human activity. Even though we can trace Freytag's sources to earlier sources like Aristotle, who defined plot as a beginning, middle, and end with the arrangement of incidents to create tragedy, the real source comes from our ancestors. We know that early humans painted cave walls as a way of preserving events for others. Telling stories is one of our essential human characteristics. Of course, it's extremely important to note that Freytag's triangle is just one storytelling structure that we use here only for illustration and as a means of introduction for students just beginning to write creatively. Many other cultures adopt other storytelling structures that are as powerful as the triangle if not significantly more so, and Jane Alison does a wonderful

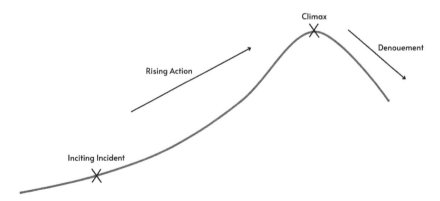

Figure 4 Freytag's triangle.

job of laying many of these out in *Meander Spiral Explode* as does Matthew Salesses in *Craft in the Real World*.

Stories—no matter what structure you as a writer employ—are often what hold video games together. How those stories are told (plot) can make it a game that you toss aside or one that you'll come back to over and over again.

Plot Basics

Writer Darrell Spencer said that there were only two plots: somebody comes to town and somebody leaves. We have since tried to locate the source. Most attribute it to John Gardner's posthumous *The Art of Fiction* but the quote doesn't appear there directly. Others say Gardner used the anecdote in craft talks, saying more specifically "A stranger comes to town" and "A man goes on a journey." While many plots diverge from these two, these two simple structures do resonate for the vast majority of plotted stories. But there are other problems with the anecdote. The two plots imply a Western-genre-like atmosphere, a guns-blazing kind of plot. Its masculine overtones and the elided woman in the second plot (pronouns matter!) have prompted some to decry their limitations. But at the heart of the anecdote is a grain of truth: plot is about change and conflict.

After you've identified a basic plot structure, you need to pay careful attention to implement it effectively in a story. Somebody goes on a journey. You have a goal and conflict. Then what? On the surface, the journey looks easy. You're on a quest! We're Frodo going to destroy the ring! But without some structure, a hero on a quest can quickly become a wandering purposeless nomad. Even most sandbox games will have goals or increasingly difficult obstacles that will make the journey of change feel earned. So here are a few basic plot shapes to help you craft compelling games.

The Three-Act Structure

One of the most common plot shapes for video games is the three-act structure. This structure takes its cues from Aristotle's basic beginning-middle-end and expands it. Each section of the plotted story becomes an act with its own conventions. Hollywood films have adopted this structure so completely that you can almost predict down to the minute when certain events will occur. They follow the three-act structure to a fault. Most Hollywood screenplays, for example, are 90-120 pages long with each page representing approximately 1 minute of screen time. The first act will generally be a quarter of the pages (20-30), the second a half (45-60), and the third a quarter again (20-30). These conventions have become so rigid in Hollywood that major plot points or transitions between structures have become predictable. Next time you're in the theater, check your watch and you'll have a good idea when you reach the first major turning point or when to expect the climax.

In video games, the three-act structure is much more malleable. The beginning act still will contain an inciting incident and a turning point, but the amount of time it takes isn't going to be measured down to the minute. The middle act will often be much longer and episodic, comprising the bulk of the gameplay. This episodic structure, sometimes expressed as levels, are used by many games in the industry, stretching the second act to hours and hours of gameplay, depending on the length, number, and difficulty of the episodes. The climax and denouement will also be shorter. But the three-act structure is still a useful way to understand how to make a story engaging for the player and to create obstacles for the character before she reaches her goals.

Act One

In video games, Act One usually comprises an inciting incident, a major turning point, and the introduction of a concrete goal. In many games this is often called the tutorial. It's where the player will learn about not only the character and central conflict but also the basic game mechanics. The first act will often rely on cutscenes to provide exposition, making the first act one of the most essential for a writer. The inciting incident, character, and character goals need to be communicated in an engaging way so that the player is invested in the game. Cutscenes need to convey game essentials and not burden the player with unnecessary information. If they don't, the player will lose interest and be more likely to skip.

An example of a compelling first act is *Horizon Zero Dawn*. A visually stunning postapocalyptic game, *HZD* succeeds narratively because of its first act. The first cutscene starts immediately during the opening credits, introducing a bearded leader named Rost carrying an infant girl on their way to a naming ritual. "Today I speak your name, girl. But will the goddess speak it back?" This question lingers as we follow the pair across a snowy landscape up to the top of a hill where they encounter an elderly woman who is there to bless the naming. The ceremony is interrupted by her peers: other matriarchs who disagree with her decision. The man and the girl are outcasts, the girl motherless, and we learn that Rost has been entrusted with her care. Still, he speaks her name, "Aloy!"

The scene conveys expository information in a short period of time while also posing several questions that the player wants answered. Who is Aloy? Why is she an outcast? What are the machines that roam the world? Where did they come from? And most importantly, what happened? The first act helps us answer many of these questions and sets Aloy on her quest. After the naming, Aloy is a young girl, curious about her identity. Once, she picks berries with some other children and then presents them to their mother who refuses her gift. Upset, Aloy falls down a ravine and finds herself in a cave rife with foreign technology that helps us piece together the past.

As we play as Aloy, we gather information and become more familiar with the controls. When she retrieves a triangular earpiece, we're introduced to a technological interface that helps us access even more information about her environment. We get videos from deceased ancestors and learn what happened—how technology took over and created the postapocalyptic world that we now live in.

After getting out of the cave, Aloy then learns from Rost how to hunt machines and how to gather resources to help her on her journey. The scene culminates with a challenge that allows her to use the new technology she has found and the skills she has acquired. The interface predicts the path of several dinosaur-like machines as she sneaks through their pack to help a comrade who has fallen and broken his leg. After she successfully rescues him, she then encounters their elders who berate her for even talking to him. She's an outcast! But she wants to learn why. They refuse. The only way for her to know is to compete in the Proving, a coming-of-age competition. Now we have a definite goal: win the Proving and we'll find out who Aloy really is.

The first act succeeds because it conveys expository information through gameplay. Even the mechanics of using the earpiece are integrated into the story. Because the conflict is tied to Aloy's character, the ensuing journey and challenges are a natural extension of the first act. We're ready to go on a journey with Aloy and motivated to help uncover the truth behind her identity.

Act Two

In video games, Act Two comprises the bulk of the gameplay, leading up to the game's climax. Unlike a movie, where the amount of time is carefully parceled out between the acts, a video game often stretches this act by using chapters or levels to extend and increase the challenge of overcoming obstacles to reach the end goal(s). Throughout this act, the player becomes more proficient, gaining rewards for defeating bosses or solving puzzles that will help the player along her journey.

Game writers and designers, then, need to be thinking about scaffolding challenges so that the player continues to progress. Make the challenges too difficult, and the player is likely to get frustrated or lose interest. Make them too easy and rewards seem hollow or meaningless. Even more important is that the challenges build upon each other naturally so that arrival at the climax seems inevitable and not gratuitous.

Act Two in *HZD*, for example, is the longest, comprising about forty or more hours of gameplay. To organize and deepen the rise in action, *HZD* divides this act into several levels or quests. After the tutorial (two quests), there are twenty more main quests. Each has their own distinct goals and story line, often organized like miniature linear stories themselves. Finishing the level/quest results in rewards and personal progression. Aloy accumulates experience points, skill points, and other rewards. But most importantly she accumulates more self-awareness and knowledge about her place in the world.

With these two plotted structures working in tandem (the smaller goal-oriented quests and the larger rising action of the second act), *HZD* effectively engages the player in a complex interactive narrative, infusing the forty plus hours of gameplay with purpose and enjoyment. In order to do this effectively, the game must scaffold tasks that enable Aloy to unlock more mysteries about the world and her place in it. Some of these tasks are more functional, while others are knowledge-based. For example, in the short-but-pivotal "Womb of the Mountain" quest, Aloy awakens inside a mountain after being saved by Rost. She has just competed in the Proving which was interrupted by a cultist attack with her as one of the primary targets. The promised knowledge about her identity from the first act is delayed to this moment. Aloy scans technology left behind by a fallen cultist and discovers information about their mission to kill her. The most disconcerting evidence is a video of a woman who bears a strong resemblance to Aloy.

One of the matriarchs comes by and compounds the mystery. Aloy wasn't born of a mother but was birthed by the mountain itself. So, who is she? And why would the cultists want to kill her? Rost taught her to use her personal gifts in the service of her people, the Nora. Her identity holds the key to the Nora's survival. The matriarch then anoints Aloy a Seeker, a Nora emissary free to go beyond their lands. At the end of the quest, Aloy must prove herself once again to a war chief, destroying an ancient machine called a Corruptor that has the ability to turn more benign machines hostile. In defeating the Corruptor, she salvages its override component which now allows her to hack into other machines and control them.

The quest then delivers tangible rewards that are useful and necessary to completing the game. The knowledge that she gains deepens her commitment to completing the game and accomplishing its goals. By scaffolding goals through quests in the second act, *HZD* effectively seeds the character's development and delays realization of the core game goals until the climax of the third act.

Act Three

In video games, the third act is often the shortest. In the third act, the player arrives at the climax and often—especially in large-scale commercial video games—defeats the final boss or solves the last puzzle. The goal in the third act is to write something that deepens the satisfaction of completing the game while effectively closing any loose ends. This is the resolution or denouement. This more complicated French word for resolution actually provides another way of thinking about plot. "Denouement" is a compound word, de+noue+ment. *Noeu* is a knot. That is what a good plot does—it knots together various strands of a narrative. By the climax, it's tight and constricted, a complex knot brought about by compounding choices that have reached a breaking point. The denouement is the unknotting or the unraveling of the knot.

Act Three of *HZD* begins shortly after Aloy learns her true identity: she is the genetic clone of one of the old world's top scientists, the creator of GAIA, an AI system that kept the world in check before a virus threatened to corrupt it. Aloy struggles with this information. So, she doesn't have a mother? She was just created as an insurance policy for destroying the HADES virus for good? But she soon realizes her importance. As a character who was initially concerned only about her own well-being, about acceptance in a society bent on rejecting her, she now realizes that her genetic makeup contains the key to humanity's survival.

Now Aloy must obtain a weapon to destroy HADES: a lance fitted with the Master Override. Once Aloy obtains this weapon, she unites with friends and forces at the Spire, to hold off HADES and its armies of corrupted machines from uploading the extinction protocol to wipe out all life on Earth. The battle is intense, injuring Aloy in the process while HADES reaches the Spire and begins uploading. A friend finds Aloy among the wounded, and she rallies together with those willing to attack the Spire. In the final battle, Aloy defeats multiple corrupted machines and drives the Master Override lance through HADES' eye, killing it. The Master Override shuts down the virus, sparing humanity from extinction.

In the denouement, the story unknots several story lines and allows Aloy to visit the grave of her genetic predecessor, Dr. Elisabeth Sobeck. This provides some personal closure for Aloy. She has grown and changed throughout the story of *HZD*. Once young, helpless, and an outcast, she has now become a leader of her people, whose concern is the well-being of others and the planet itself, a worldview that is similar to the woman who created her: Dr. Sobeck.

Although we've oversimplified the complex story of *HZD* here, focusing almost exclusively on the narrative of Aloy's development, the game shows how a three-act structure can work to provide an umbrella framework for a game. It introduces a central conflict, develops a story line through increasingly difficult obstacles, and arrives at a climax that builds naturally from the rise in action. The structure helps the writer or writers to compartmentalize when to introduce conflict and when to build tension. Other frameworks exist with similar structures. A five-act structure has slightly different demarcations for when certain acts begin and end. The first act is the prologue or exposition, leading up to the first major turning point, not too different from the first act in a three-act structure. After that, the structures diverge with distinct acts for rising action, climax, falling action, and denouement. Five-act structures work particularly well when a writer needs more checkpoints for plot and character development.

But neither a three-act structure nor a five-act structure should be a replacement for a natural progression of a character working toward an end goal. Some narrative designers actually discourage writers from adhering too strictly to these structures in games. At the 2014 Game Development Conference, for example, narrative designers Tom Abernath from Riot Games and Richard Rouse II from Microsoft Game Studios headlined a panel titled "Death to the 3-Act Structure." They researched how gamers actually proceeded through narrative games and found that only 47 percent of players finished narrative open-world games. This meant that often players would explore side quests or develop their character in other ways rather than following through on the main narrative. "Games are not movies," they said. And while they acknowledged that a three-act structure could be helpful for organizing or thinking about turning points in a narrative game, it shouldn't be used as the yardstick for success. A better model, they argued, was a serialized model, similar to Netflix shows you binge on TV. These games had distinct narrative arcs with their own mini inciting incidents, climaxes, and denouements that were focused on particular characters. They also better matched what gamers found truly memorable about narrative games. When surveyed, narrative gamers highlighted investment in characters over identifying plot. In other words, what's most important in narrative games is similar to what's important in all good stories: character and progression. Structures have their uses, but they are secondary to developing characters. If a story becomes too plotted or concerned about reversals and betrayals, the gamer can feel manipulated.

Branching Narratives

The interactivity of video games has also complicated plot and shape in narrative, often disrupting a linear story. If we return to the most basic plot: someone comes to town and somebody leaves; interactivity can expand what happens on the journey so the journey changes based on the choices the player makes. Although choice may still push the narrative into a linear progression, the variation in plot can expand the narrative in multiple directions, like a tree trunk with multiple branches.

Branching narratives have become common in AAA games as well as in text-based indie games using tools such as Twine or Ren'Py, which we introduce how to use in Chapter 12 ("Tool Suites and the Game Industry"). Choice can deepen the player's identification with the playable character and increase replay. Most games now will have some level of player choice, allowing players to customize characters or to alter the story line based on their interests. Having more autonomy increases players' investment in a character and hence the game. Branching narratives create plot choices that allow the player to better shape their experience.

A branching narrative plot may look something like this (Figure 5):

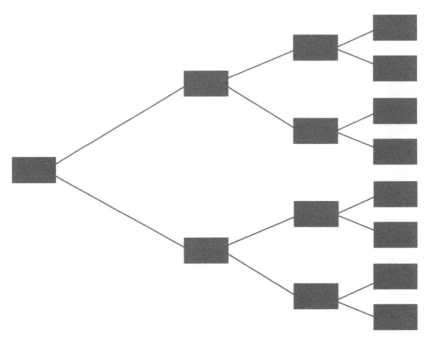

Figure 5 A branching narrative plot.

But incorporating multiple choices doesn't mean that a writer must entirely relinquish an attempt at structure. Some games allow for branching but force choice back through major turning points in a story. Dotnod's recent *Life Is Strange: True Colors*, for example, still maintains a traditional three-act structure. The main character Alex Chen first arrives in Haven Springs to live with her brother Gabe. Each step toward the end of the first act is divided up into episodic chapters with their own tasks and goals: getting to know Gabe's girlfriend Charlotte and her son Ethan, finding a job, learning about the town's mining industry, then finally saving Ethan who has gotten lost in the mines. All the tasks move us closer to Gabe's death, the turning point that sets the stage for the second act: to grieve and then find out who is responsible.

Within that structure, the player can make multiple choices that affect the outcome of the game. The player can choose between two love interests, for example: a young woman musician who runs a record store or the park ranger who is Gabe's best friend. The third act has a different denouement based on which relationship you've chosen to pursue. The player can also choose how to use her powers, a kind of super empathy that allows her to assume the emotions of others. At a pivotal point in the game, Alex confronts Charlotte. Charlotte's feelings are complicated. On the one hand, she loves her son Ethan and is happy that Gabe saved him at the mine. But she also feels anger and depression. Her son is the reason Gabe is dead. If only he had listened to her and not gone to the mine in the first place! The emotions are almost too much for her to bear. Alex realizes that she can help her—she can use her powers to take her pain and get rid of it so that Charlotte can move on. Or she can allow her to work through the emotions on her own. The choice affects the outcome of the game dramatically, ending with a character who has a healthy emotional attachment to Ethan and others around her or a cipher who seems distant and withdrawn.

So instead of a simple branching narrative structure, *True Colors* looks a little like this (Figure 6):

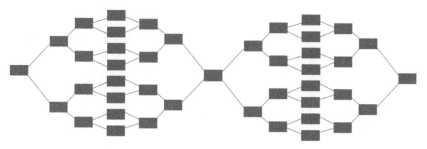

Figure 6 A more complex branching narrative plot.

This plot structure may be the most useful if you're writing in Twine, one of the tool suites we use most frequently in our classes. One of the problems we've often had in Twine is that students become overwhelmed with choice, causing the game's story line to expand almost exponentially. If you're writing the game collaboratively, the problem can become even more acute, as writers may pursue some story lines and neglect others without consideration for the game as a whole. That's why thinking about acts with specific goals, turning points, and narrative events can help you ground your choices in a larger overall structure. Characters can make choices, but only some choices. They'll still have to funnel those choices through pre-identified plot points.

There's another, more coarse, reason that choices are sometimes so limited in game development: it's cheaper. Creating games at the commercial level is incredibly expensive. The more cutscenes and the more variety of choice, the larger the narrative becomes. Each additional line requires writers, actors, designers, animators, and so forth, to make those sections of the games playable. Technology can also limit the size of a game, making a truly expansive branching narrative too large to store. Even so, today's games can accommodate significant capacity for multiple narrative lines. *Dragon Age: Origins*, for example, utilized a script with over 740,000 words. That's a massive undertaking, a choose-your-own-adventure novel on steroids.

As you grapple with shapes that expand choice or limit them, other more specific shapes may emerge. Sam Kabo Ashwell on the *These Heterogenous Tasks* blog writes about several standard patterns in choice-based games. As some games force more choices through bottleneck plot points, they take on other shapes. A more linear story becomes a Gauntlet, where choices are limited to short side quests that rejoin the main story line or diversions that result in failure, forcing the player to backtrack. Stories with more choice can become Quests, where the player travels to different locations with a set of branching choices at each one. Other models—the Spoke and Hub or Loop and Grow—deal with choice and bottleneck in different ways. A Hub can be a central place or a set of nodes. Exploration leads out from this place and then back. Loop and Grow circles around the same set of options multiple times, unlocking various possibilities as the playable character grows and progresses. All these models use the mechanics of time and choice to plot out a series of events. The only one that doesn't is the Open Map, the shape used in the majority of what the industry calls sandbox games.

Sandbox Games (*Skyrim* vs. *No Man's Sky*)

So what about these Open Map or sandbox games that don't seem to have a plot? As in the example of *No Man's Sky*, we don't have any backstory, no linear expectations. We're there to explore the world(s) and not much else. Is there even a need for a plot?

The short answer is yes. If we go back to our original definition of plot as a deliberate sequencing of events to create an impact, it might be difficult to see how this would be the case. Sandbox games allow characters to explore an environment often without entirely clear end goals. They're not linear but recursive. Still, they do promote change and development. As the player explores, the player grows. They learn. They accrue knowledge, artifacts, and abilities. And while a sandbox game might eschew plotted turning points, they do have markers of progression that can deepen emotional investment in a game. Often the imposition of these kinds of plot-based elements in a sandbox game will actually increase the game's success, giving an added purpose to all that nomadic exploration.

Take the ever-popular *Minecraft*, for example. This sandbox game has evolved into the best-selling video game of all time. The blocky graphics and simple mechanics have spawned their own culture, now with annual *Minecraft* festivals and users topping 100 million. Much of its success is thanks to its adaptability and applicability. It's a game that can be an educational tool, a multiplayer battle royale, or a role-playing game with user-generated adventures. It even recently incorporated a story mode.

Does it have a plot? Well, that depends. What *Minecraft* has done, and what many sandbox games also do, is create the tools for people to write stories of their own, sometimes with rudimentary, and sometimes complex, plots. The multiple modes (survival, survival hardcore, creative, adventure, peaceful) allow users to enter the Minecraft world where they feel most comfortable. If they're a beginner (many of the twelve-year-old early adopters in 2009 are monied VIP content creators now) they can start in peaceful mode and concentrate on mining, building, and learning game mechanics without worry of being attacked by zombies or other players. The most simple plot is to hone your virtual plot of land, a little like tending your own garden. As the player grows in ability, they'll branch out, trying servers with more plotted story lines. The largest server, Hypixel, has multiple games, including competitions with other players to develop a world in a certain amount of time. At the end, users will vote on the most successful, a little like *American Idol* but with other *Minecraft* aficionados as the arbiters of quality. In *Minecraft Classic*, we have the central *Minecraft* plot: build your inventory, craft the proper tools, put them in order to create a portal to the Ender dragon, disable the Ender dragon's ability to regenerate by destroying the crystals on top of the circle of Stonehenge-like obsidian monoliths, and then kill the Ender dragon. The result leads to a short resolution that plays the game credits till the player is invited to return to the world and do it again. So, plot does exist in *Minecraft*. Players can go on quests, battle with other players, complete timed trials, and create their own stories. It's a place where one can apply the rules of good storytelling from the most simplistic to the more advanced.

The other place that plot exists in sandbox games is in multiplayer environments where the story develops organically in the interactivity between players. Sometimes this environment is a competition between other individuals, sometimes it's in teams. The story emerges as the players work together or against each other. Recording the adventures and reposting them on YouTube immortalize these stories, becoming records of heroics in a virtual environment that aren't all that different from our ancestors etching stories on cave walls.

Case Study: *Fallout 4* (2015)

The story of *Fallout 4* begins in the year 2077 with society on the verge of nuclear war. The player assumes the role of either Nate or Nora, a married couple raising a baby named Shaun, who makes the decision to purchase cryogenic tubes in an underground vault where they can be frozen until the world is safe. Years pass, and their vault is broken into by two strangers who murder the player's spouse and kidnap their baby. The player's cryogenic chamber eventually malfunctions, and they're now free to explore the radioactive world of *Fallout 4* with a singular mission—to rescue their son Shaun.

When Sal assigns *Fallout 4* and asks students to share what their Nate or Nora embarked upon, their answers are always extremely varied. Some of them spend their time building basketball courts. Some rig elaborate hideouts complete with electricity and pinball machines. Some explore the USS Constitution and chat up its many robots. Some join the Brotherhood of Steel and patrol the land in mech suits. And some just bum around Boston, exploring the devastated city. Very few follow the intended story in a straight line and attempt to save Shaun.

And yet, does this make any kind of narrative sense? The opening cutscene characterizes Nate and Nora as loving parents who would do anything for their child. Why would they suddenly decide to build basketball courts after he's kidnapped, while some players choose to ignore Shaun altogether?

In academia, we've defined the frame narrative in games as any moment when the player is stripped of control so a story can play out. Conversely, any time the player has control is the ludonarrative. An easy way to remember this is understanding that "ludo" is Latin for play. Therefore, the ludonarrative is simply the narrative of your play. The example from *Fallout 4* is a case of ludonarrative dissonance: a moment or series of them when the ludonarrative comes into direct conflict with the frame narrative. Tom Bissell, in his book *Extra Lives*, describes ludonarrative dissonance by highlighting a mission in a *Call of Duty* game, where the player must infiltrate an enemy base with their partner. The player needs their partner to survive, and there's no way the game can progress without them. However, if a mischievous player chooses, they can turn their rifle on their partner and execute them. It makes no sense in the story whatsoever, and this is what we as game writers must try to avoid—creating stories that can so easily be torn to shreds by players (Figure 7).

Open-world games are especially vulnerable to ludonarrative dissonance because there is simply so much player freedom and choice. How can any writer twist hundreds of player choices into a logical story? Compare the opening of *Fallout 4*—which almost begs the player to break its narrative by quickly providing them with dozens of activities that have nothing to do with rescuing Shaun—with the start of *Skyrim*. In that game, the player assumes control after an unnamed character is captured for illegally crossing a border and subsequently freed during an execution gone wrong. However, the player creates their character and their motivations. No matter what they choose, their choices inherently make sense for that character, because the player collaborates in creating the protagonist, mentally justifying their decisions.

Skyrim, despite being extremely open-ended, rarely has moments of ludonarrative dissonance, and it's instructive to consider why. In this genre, the counterintuitive move of leaving certain beats of characterization or motivation blank actually contributes to a more believable and fleshed-out story. Ceding that narrative control to the player is often vital in creating an open-world game with an immersive

Figure 7 *Fallout 4* revels in its many dioramas of postapocalyptic decay.

and cohesive story. In this genre, writers must collaborate with the player on the story.

One final note we always tell our students about ludonarrative dissonance is that sometimes those dissonant moments are the most "fun" aspects of the game. "Fun" is impossible to quantify because it's so singular and personal, but let's explore this through the lens of *Grand Theft Auto: Vice City*. When Sal originally played this game in 2002, the most fun he had was stealing tanks and blowing up police. This had nothing to do with any missions and contradicted the story where protagonist Tommy Vercetti—voiced by *GoodFellas* alum Ray Liotta—is tasked with uncovering why a drug deal went wrong. Why focus on killing cops when you're supposed to be solving a gangland mystery? This balance between ludonarrative dissonance being "fun" while simultaneously disrupting the narrative is something all game writers should keep in mind, and even Rockstar has tried to remedy this to some degree. *Grand Theft Auto V* puts you in control of Trevor, a character so unhinged and violent that you absolutely could believe that he would take a break from the game's overarching narrative to partake in a meaningless murder spree.

Adaptability to the Creative Writing Classroom

A few years ago, after Eric started teaching writing for video games in the creative writing classroom, he talked to a colleague about some of the ground rules he set for a fiction workshop. Students were welcome to work on whatever they wanted but they couldn't write anything that he called "genre fiction." No elves. When Eric asked for his reasoning, he said that he was simply interested in good storytelling. Students would learn the rudiments of language, character, and plot and then they could apply those principles to whatever genre they wanted to. But couldn't they learn the rudiments of fiction while also applying them to a specific genre? What did he have against elves? Other reasons emerged: most often the stories were longer, and students were more focused on worldbuilding than story. Then the final reason: as someone who never read much genre fiction, he felt ill-equipped to guide them.

Eric asked for an example. One student begged to work on a high fantasy novel that she'd been writing since high school. Then, for her first workshop, she turned in fifty pages of lore. Fifty pages of backstory, history of the four kingdoms, the reason dragons were restricted to nobility. Fifty pages of character sketches, descriptions of races, the economic and sociopolitical morass of this student's own Middle Earth. Eric's colleague told her to hand something else in.

This harsh attitude is not uncommon in some creative writing classrooms. Ben Percy writes extensively about his own experiences writing "genre fiction." In his first fiction writing class as a graduate student, his professor told him: "no genre." When he asked for clarification, his professor said, "No vampires, no dragons, no robots with laser eyes." Percy was exasperated and still didn't understand. He asked, "What else is there?" Writers of his generation have interrogated biases for the kind of writing his professor was advocating for—let's call it "literary fiction." Other writers like Carmen Maria Machado in *Her Body and Other Parties*, Colson Whitehead in *Zone One*, and Marlon James in *Black Leopard, Red Wolf*, to list a very small few, have carved out spaces where genre elements are combined with language or character-driven stories that "bend genre." Instead of creating arbitrary designations and dismissing genre, these writers advocate welcoming multiple literary traditions into the workshop and breaking down false literary/commercial hierarchies. Could there be a way to guide Eric's colleague's student so that her fifty pages of lore could better serve her story?

In an open-world sandbox game, knowing the world is important so that players can have an experience adapted to their interests. Some players become enamored with a given world and game and want to spend more time in it. Some open-world RPGs like *Skyrim*, for example, have made an industry on side quests and exploration beyond the main story line. The quality of the lore, the variety of tasks and interest in completing them, or the ability to make stories yourself increase the replayability of the game.

Another way to guide the student then would be to look at the quality of the lore and see if it lends itself to the player doing something within it. What are some potential goals? What will the character do? We often have students who struggle in fiction with plot. They have understood the concepts of showing, not telling. They've learned to describe using concrete language and imagery, to write characters that are complex and interesting with dialogue that pops. But sometimes the student will get stuck in a scene where not much happens. They describe a coffee table littered with mashed tissues and a sepia-toned couch. They walk into the kitchen and clack open cupboard doors looking for something to eat. The description starts to overwhelm the story, and in so doing, begins to bore the reader.

One way to think about plotting a video game is to put action on an axis. In a sandbox game, the game moves more horizontally. The player can go into a room and start opening cupboards. Why not? They're exploring. But at a certain point the character will need to find something in the cupboards that leads them to do something else. In other words, they need vertical action. The ability to make things happen in a video game, interact with an environment, uncover goals, solve puzzles, or otherwise grow and develop a PC, provides meaning to a player. The key in looking at an open-world sandbox game is asking: Is there something to do in it? A topographical map of action in such a game might look something like Figure 8.

A successful game needs to have these moments of plot. They might not be constricted to a linear three-act structure but they exist as the blips of action in a horizontal playing field. If it doesn't, the player will become bored. What am I supposed to be doing? And eventually lose interest. If they find things to do, well, then you may have a player for life.

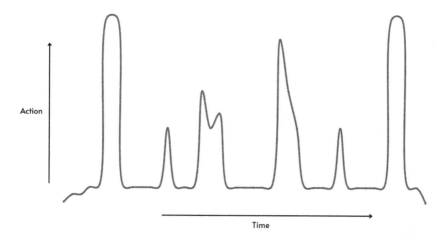

Figure 8 A topographical map of action in a video game.

So, *No Man's Sky*.

Why was it initially less successful? Conceptually the game was a hit. *Minecraft* in space! An open-world game of space exploration like you've commissioned your own Starship Enterprise off to boldly go where no one has gone before, never mind the split infinitive! Hello Games had designed a compelling concept with algorithms that continually grew the sandbox. But what were gamers to do with all that space? Planets are big! And far apart! Combine that with the limited capabilities of early ships and players were stuck for sometimes hours just to move from place to place. Luckily, Hello Games acknowledged the problems with the first iteration of the game and worked on it to make it better. The money that they made off the concept went back into development. Updates added what was sorely missing before: a plot.

No Man's Sky today still retains its roots as an intergalactic sandbox game but now gamers benefit from an array of plot-based tasks. As early as six months after the game's release, Hello Games started introducing updates. Some were for improving the mechanics of the game, allowing for more customization or the ability to build bases, tame creatures, or colonize new worlds. But the most significant updates concentrated on plot: introducing conflict and developing an overall story arc. In one of the early updates, players receive a distress call from Artemis, a member of a traveler race who is in need of help. Artemis directs the player to triangulate their position, leading to eventual contact. The introduction of this character leads to tasks that are now laden with purpose. No longer mining for resources to merely improve a ship, the player now must complete missions in order to help someone and uncover the secrets of the universe. They're also introduced to antagonists that will impede their progress: mechanical sentinels who police the universe and can be extremely hostile. Although players are still free to explore the world of *No Man's Sky* without engaging the various plotted story lines, the added features give another purpose to players. They can help save the universe. The current version of *No Man's Sky* has multiple levels in a loosely plotted structure, making it one of the most exciting open-world sandbox games on the market today.

Overview

In this chapter, we've introduced the concepts of plot and story. Story is *what* happens and plot is *how* it happens, deliberately constructed for dramatic effect. We've outlined some structures to help organize and plot your stories and shown how plot functions in games. Done effectively, plot can increase player engagement and entice the player to come back to a game again and again. In the next chapter, we'll look at building the worlds where your stories take place.

5

Setting and Worldbuilding

You're in an elevator-sized submarine that looks like something out of a Jules Verne novel: metal rivets, Art Deco interior, and a simple control pedestal with a lever. You're still on an adrenaline high from the circumstances that brought you here: the plane crash, swimming through a minefield of debris. Then this odd lighthouse emerged from the gloomy dark. Every moment you swam and spit water outand climbed or descended stone stairs has been filled with a feeling of predestination. And now you're here in this metal contraption with a lever waiting for your hand. What the hell. You pull it.

The doors close. The submarine pressurizes. A wall of bubbles streams up. Signs in Aviator font mark the depth: 10 fathoms, 18. You pass a statue of a diver, hands held high. The features are linear, simplified. A screen pulls down in front of you and a static-image newsreel begins running. First, an advertisement: Plasmids! Fire at your fingertips! A man lights a woman's cigarette using what? His hands? Next, a still of a suited man smoking a pipe sitting in a leather office chair.

> I am Andrew Ryan and I'm here to ask you a question: Is a man not entitled to the sweat of his brow? No, says the man in Washington. It belongs to the poor. No, says the man in the Vatican. It belongs to God. No, says the man in Moscow. It belongs to everyone. I rejected those answers. Instead, I chose something different. I chose the impossible. I chose Rapture.

You're released from the elevator shaft and your submarine floats out into the underwater vista of a submerged city. Watertight walkways connect skyscrapers. An enormous squid propels itself away from you revealing its suckered arms. Strands of seaweed provide cover for schools of fish. Flickering neon signs announce Bella Mia's High Fashion, Pharaoh's Fortune casino, hotels, and a cocktail lounge. Off in the distance, a whale swims through the buildings like it's a blimp in a dim sky.

> All good things. On this Earth. Flow. Into the City.
> This is the world of Bioshock.

Some of the most important and enduring aspects of video games are the immersive worlds that they are able to create. Like movies, video games have developed technology to effectively convey sometimes fantastical locations with realistic effects. Animators and designers spend countless hours perfecting these elements, layering the landscape with detail, highlighting the spires of a city, or coloring boulders with blue lichen on some alien world. Your job as the writer is to imagine the environment that will inspire your collaborators to sketch and design and populate the world, to give it a feeling of verisimilitude. You know whose head is stamped on brass coins and what scaled creatures haunt the world's oceans. You know the ghosts of the world's past.

Setting is more than just the backdrop for the narrative. It's an integral part of the gaming experience, used to convey a sense of time, place, and mood. Done correctly, setting can become like another character whose influence is felt across the levels of gameplay, embedded in the rules and laws of your interactive world. If you change elements of this world after carefully introducing them, you can have dissatisfied players, or players who aren't moved or compelled to come back to the game again and again. A complete world feels like another home, a home that you don't want to leave.

Like character and plot, setting is best when it resides in the particulars, when place builds from specific concrete details. This is true whether you're describing the streets of Chicago or *BioShock*'s Art Deco underwater world of Rapture. Avoid statements like "It's in Anytown, USA" or "some distant planet." These abstract placeholders show a lack of imagination. We see the intention: the writer wants to make a place open enough that anyone could come and add or contribute to it. This may work in certain contexts like for a sandbox game where the player is responsible for most of the content but with a narrative game, ironically the opposite happens. The blandness of the abstraction lends itself to cliché and generalization. If it's Anytown, USA, the streets become any streets, the stores any stores, the people any people. This lack of marking can lead some writers to maintaining a status quo that is often white, male, middle class, and heteronormative. Create a world with more specifics and that world is more likely to be inclusive, the kind of place where diverse players will come to play.

So if you start with a distressed mining spaceship (*Dead Space*) or the post-Communist/post-Capitalist urban decay of Revachol (*Disco Elysium*) or a lake-side sunset in the drought-ridden Shoshone National Forest (*Firewatch*), then you have a place to begin your game, a place that you'll want to interact with, a place that will activate the imaginations of the people you're collaborating with to design buildings and landscapes and gameplay beyond what you could have produced yourself.

Managing Time and Place

One of the most essential tasks of a writer is to convey the setting of their world quickly and effectively. The audience needs to know where and when the story is taking place. Without markers to ground the player's experience, the player feels adrift and lost. Ambiguity is disorienting. Information about when and where a game takes place is necessary for the player's immersion. You need them to help the player take the plunge.

If you're working in a primarily text-based medium like Twine, you can communicate setting through description. Prose writers have long mastered describing new and strange settings using economical language to convey both time and place. Consider this description from Sequoia Nagamatsu's *How High We Go in the Dark*, a novel about a global pandemic so severe that the government has established euthanasia parks where parents can bring their children to die. Here, a first-person narrator describes seeing the park on his first day on the job:

> A teenage girl named Molly wearing pink striped overalls escorted me to the employee dormitories. Beyond the old prison buildings, the park resembled a Six Flags knockoff—cracked pavement, kiosks filled with off-

brand candy, papier-mâché dragons and enchanted fairy forests that looked like there were about to melt in the sun or dissolve with the next downpour. The central prison complex had been renovated into a pirate-themed shopping area and food court inappropriately named Dead Man's Cove with vendors and vending machines and food carts and animatronic displays occupying the cells. Above us, rainbow spotlights scanned the ground from the guard towers. I could see the silhouettes of their rifles amid the prismatic glow.

Nagamatsu contrasts the "Six Flags knockoff" and the prison remnants, creating an eerie setting. The details designed to elicit amusement-park-like fun such as the animatronic displays or candy-filled kiosks take on the weight of children's deaths. The setting is effective because it doesn't serve as merely a static backdrop for the action in the scene. Every detail of the park is saturated with potential energy: the off-brand candy and papier-mâché that look about ready to dissolve or the menacing guard towers with their rainbow spotlights. We know at some point that these guard towers will activate, despite the girl Molly's reassurance later that "it's mostly for show." The whole park exudes impermanence and death, from the dissolving candy to the refurbished cells in Dead Man's Cove.

Textual descriptions in video games can similarly activate the setting so that it's not just a backdrop for the game. The game *Lifeline* from the studio 3 Minute Games uses the setting of a distant hostile planet where a lone astronaut Taylor has crash-landed. Players receive text messages from Taylor in real time as they navigate the planet, trying to survive:

All right. So my escape pod came down in some kind of desert.

The ground is all cracked white rock. There's a huge white peak a few miles away.

It's weirdly symmetrical, like it might not be a natural formation.

My IEVA suit's compass places the peak northeast, and then, in the opposite direction, south and southeast, to be precise . . .

. . . are two funnels of black smoke from what I have to assume are two pieces of the Varia.

Best case scenario it's ONLY in two pieces.

The crash site looks closer than the peak. What do you think I should do?

Each description here implies potential action. The "huge white peak" or the "two funnels of black smoke" represent two distinct destinations, two beginnings for survival gameplay. They're not simply static descriptions of Taylor's world but also indications of what the character might do.

Sometimes setting is conveyed through shorthand descriptions in titles or chapter headings. A text-based game might opt for this solution in lieu of paragraphs of description. This has become a staple of historical novels. Georgia Hunter's *NYT* best-selling novel *We Were the Lucky Ones*, for example, starts each chapter with a name, location, and date: Addy, Paris, France, early March 1939. While this technique is economical, it's also artificial. We don't know how or where we get this information except from some all-knowing omniscient author.

Visual video games may similarly rely on the economic use of titles or headings if visual or textual cues aren't enough. *BioShock*, for instance, starts with a black screen with the words "1960: Mid-Atlantic" superimposed over it. This orients us time-wise so that we don't question elements like smoking in the plane cabin or the Art Deco architecture or other stylistic elements. The developers still likely could have done without this information, since so much of the time and place is conveyed through gameplay but the shorthand title helps the player get quickly oriented. But the rule is the same as traditional texts. You can usually do without. Visual cues can be enough without relying on a title screen that can sometimes ruin the dream or draw too much attention to itself.

Setting as Mood and Tone

Setting is also a powerful tool for communicating mood and tone. Tone is the writer's attitude toward the setting. Combined with the chosen setting, tone can create an overall feeling, atmosphere, or mood. In traditional print media, descriptive language is the primary medium for this. Most of the time, the setting and the narrator's tone will be complementary. The narrator finds herself observing a panoramic vista, and her language reflects her attitude of awe and feeling of insignificance. But writers will often contrast the tone with the setting, sometimes to profound effect. A setting that is desolate can be transformed with a lyrical narrator.

In film or video games, tone and setting are often communicated visually, the decisions left to teams of animators and designers. Most AAA video games will align tone with setting. Survival horror will feature dimly lit corridors and nightmare monsters, usually with the protagonist trapped in some morbid location from where there is no escape, like a giant mining spaceship (*Dead Space*), a dilapidated psychiatric hospital (*Outlast*), or a twisted forest cut off from the rest of the world (*Darkwood*). A desolate setting will convey a desolate tone.

But sometimes contrasting tone and setting can deepen the poignancy of a game and its artistic cachet. *Bioshock Infinite*, for example, uses a fictional floating city called Columbia that looks like the 1893 World's Columbian Exposition in Chicago. The bright colors, stately monuments, and park-like atmosphere starkly contrast the gloomy underwater world of the original *BioShock*. But the tone is just as sinister.

Case Study: *Final Fantasy VI* (1994)

When it comes to setting, *Final Fantasy VI* remains a fundamental example of how to produce a vivid and changing world in the medium of video games. Cowritten by Yoshinori Kitase and Hironobu Sakaguchi, *Final Fantasy VI* begins like so many RPGs—a mysterious hero with amnesia realizes they're being manipulated by an evil Empire and joins up with a large and colorful cast of rebels to take them down and save the world. This Super Nintendo game then diverges from what, in 1994, had become a familiar template. The defacto protagonist of *Final Fantasy VI* isn't some silent male knight like in the original *Dragon Quest*. Instead, we're given control of Terra, an eighteen-year-old woman with an elaborate backstory that trickles out over thirty to sixty hours of story. Although familiar knights and wizards and dragons appear throughout *Final Fantasy VI*, they're dropped into a steampunk setting where the devious Empire uses the calcified corpses of magical creatures to fuel everything from airships to roving robots that wouldn't look out of place in an anime like *Neon Genesis Evangelion*. Events proceed according to the template from there, but a strange thing happens on your way to saving the planet: you lose.

Midway through the game, Terra and company confront Emperor Gestahl atop the Floating Continent high above the ocean—a climatic encounter that feels like the end of the game. In a shocking twist, the Emperor's cackling underling—the mad clown Kefka—kills the Emperor before triggering an ancient fail-safe that effectively destroys the world. The player is given two minutes to escape the exploding Floating Continent, and if they don't reach their airship with enough time to wait for one of their comrades, he permanently dies, a branching story path that can't be undone. The game fades to black as the planet tears itself apart, and, for many of us who played *Final*

GIRL: I can't remember a thing...
OLD MAN: Don't worry.
It'll all come back to you...in
time, that is.

Figure 9 A young woman suffering from amnesia is reluctantly drawn into rebellion in *Final Fantasy VI* (1994).

Fantasy VI in 1994, it remains the most shocking moment in video game history (Figure 9).

When the game resumes, the player is given control of Celes, one of Terra's comrades, but certainly nowhere near the level of a protagonist. She's washed up on a tiny island with only one other inhabitant—Cid, her surrogate father. Neither is sure what happened to the world or even if anyone else survived. Days pass, and Cid grows sick and nearly dies. The player can't do much beyond fishing in the decaying waves. An apocalyptic red haze hangs over everything, and if you don't catch enough fish for Cid, he also permanently dies. Crushed by the prospect of spending the rest of her life alone, Celes hurls herself from some island cliffs to her death.

What is perhaps the darkest moment in any Super Nintendo game is followed by a series of downbeat reveals. Celes awakens on another shoreline and discovers that some minor form of humanity has persisted—including the devious Kefka who's installed himself as ruler of the dead world. The climate and food supply are devastated, and there's little hope for a meaningful future. Over another fifteen to thirty hours, Celes tracks down the surviving members of her team—

including Terra. But even this is optional. The player isn't required to do any of this to "finish" the game. All they have to do is storm Kefka's tower and kill him, and the game ends. When we originally played this game, we of course assumed we'd stumble across some way to fix the broken world, some magical solution to everyone's problems. However, there is no deus ex machina. No matter what you do, no matter how strong you become, there is no reassembling the world that Kefka cracked open.

That this narrative is threaded through with hope is a miracle of Super Nintendo storytelling. Its world is flat and pixelated, and the characters are barely capable of more than a handful of simple expressions. And yet, each member of this large cast of characters must find their own complicated reason for hope. *Final Fantasy VI* is a prescient story about living through societal and climate collapse, and its ambitious twists and turns still feel fresh today. It's a game worthy of study not just for those interested in writing RPGs but for anyone interested in narrative, mood, and tone in any medium.

Realism versus Fantasy

It can be difficult to imagine either realistic or fantastical worlds that will stun with their uniqueness. Students sometimes fall into traps, relying too much on convention to flesh out their world. Consider this example from an aspiring novelist who is writing a high fantasy novel about dragons. The novel is well-plotted and has an interesting enough protagonist but the world is thin. The dragons, for example, don't seem to have any origin and aren't integrated into the world except as weapons to be used in battle. They have no means of communication, no reason to be there other than to enhance the very plotted plot.

The problem here lies in the novelist's worldbuilding, in their ability to imagine a setting beyond certain surface conventions of the genre. For revision, the writer needs to defamiliarize the setting so that it is unique and new. Writer/theorist Charles Baxter talks at length about the concept of defamiliarization in his book *Burning Down the House: Essays on Fiction*. The concept comes from Russian critic Viktor Shklovsky. Defamiliarization means "to make the familiar strange, and the strange familiar." So, if you're writing a game about dragons, you need to make whatever is too familiar more nuanced or "strange." We need to know how the dragons fit into this world and what their purpose is. They need autonomy just like the human protagonists. What is going to be different about *these* dragons, this world? The game also needs to make the strange familiar, to imagine the world and its rules so wholly that it seems realistic, like characters can live and operate in this world, going about their daily lives.

This relates to one of the most common pieces of advice we hear in creative writing classrooms, which is to write what you know. Whether your world is the steampunk landscape of *Final Fantasy VI* or as mundane as your own backyard, you need to know the world, its rules and customs and textures in order to communicate them effectively to your audience. If you don't know it well, if its rules/laws don't make sense or are too general to be convincing, the player will respond by not playing the game or complain about it being cliché.

So how do you get to know your world?

The Role of Research

The advantage of choosing an actual place and time is that you have a concrete starting point for research. Sometimes AAA developers will hire historians as consultants to make sure that the details of the world are accurate. Ubisoft, the developer of the *Assassin's Creed* franchise, hires full-time historians to get the details right. This attests to the knowledge one may need to acquire to render a world effectively. If you set your game in ancient Egypt (*Assassin's Creed Origins*) or early twentieth-century wild west (*Red Dead Redemption*) or the contemporary streets of Harlem (*Miles Morales: Spider-Man*), you have a wealth of information that you can find from the books or the internet to create a realistic interactive backdrop for your game. The disadvantage? Well, these are places that other people know as well and adhering to the rules of verisimilitude can have its limits, sometimes getting in the way of good storytelling and becoming a time sink as you research more and more and don't make progress on your game narrative. How much research is enough research is always a hard question to answer—write what you know!—but a good litmus test is whether or not it's impeding or enhancing the story.

Another danger: other people know your world too and have knowledge that may exceed your own. Most fan mail that famous authors receive are from readers trying to correct inaccuracies that they have found—Charlie's isn't on J street!—and that real-world rigidity can be tiresome. Some writers have unique work-arounds for these kinds of problems. One is to choose a fictional city in a particular real time and place that is already familiar to the writer. For example, the game *Life Is Strange: True Colors* that we mentioned in Chapter 4 is set in the fictional town of Haven Springs. It borrows from other small towns in northeastern Colorado with a mining history. The main street historic downtown with its clapboard wood storefronts and sweeping mountainous vistas give the fictional town a feeling of verisimilitude. Deck Nine Studios, the studio responsible for *Life Is Strange: True Colors*, is also located near Boulder Colorado, an area similar enough to Haven Springs that the developers didn't have to do all that much research for the setting to feel authentic. Choosing a fictional town freed the studio to create the game without worrying about criticism for local inaccuracies. It's a place that the developers know and it shows.

But sometimes simply creating a fictional setting in a real-world location isn't enough, especially when you're representing a place that crosses geographical and cultural boundaries. Even the most well-intentioned and researched game can have blind spots. Take, for example, the much-hyped 2020 game *Ghost of Tsushima* developed by Sucker Punch Studios. Set in feudal Japan, *Ghost of Tsushima* took as its inspiration the actual Mongol invasion of Tsushima in 1274. Sucker Punch worked closely with Japanese historians and their team visited the Tsushima region twice to get an accurate depiction of the setting. But in development, Sucker Punch listened to some local advice and neglected others, resulting in a setting that was full of culturally insensitive fictional distortions. The samurai, for example, are not historically accurate; ignoring historians, the studio chose images of samurai from Kurosawa films, a much later period of Japanese history that was more in line with Western expectations. Similarly, the haiku side quests are completely anachronistic. Even the island itself, with its identifiable topography and actual history, was altered in favor of a fictional one by the same name.

Choosing an entirely fictional world for your game also requires research, starting with getting to know the variety of fictional worlds that already exist in your given genre. There's nothing more frustrating than spending time creating a fictional world to then realize that someone else has already dreamed up something similar. Researching requires not only a knowledge of contemporary games and trends but also a historical understanding of the genre. A good starting point for genre research is Chapter 7 of *Story Mode* ("Demystifying Genres"). Researching other games in a particular genre with an eye to creating your own fictional world usually results in opening up possibilities rather than limiting them.

Communicating the World

In visual media, you have to provide the details of the world in context. Sometimes a short establishing shot is all we need to ground the viewer. A long shot of the Eiffel Tower and we know we're in Paris. A shot of the Giza Pyramids and we're in Egypt. Many video games employ this technique. *Final Fantasy VII* begins in a similar way. Press start, and the camera lazily takes in a tapestry of stars before cutting to a close-up of the face of a young woman named Aerith. The camera pulls back, and we see that Aerith—holding a bouquet of flowers she hopes to sell—is bending over a gutter in a futuristic theater district complete with revving trucks and motorcycles. The camera zooms out even further, and we glimpse the city itself—Midgar, a steampunk dystopia where a massive power plant straddles a pizza-shaped metropolis. The rich live up top among the fresh air, and the downtrodden live below in what the script repeatedly defines as "slums." Cut closer to the power planet, and we see a group of armed militants leap from a train and knock out two guards. It's no surprise that *Final Fantasy VII* goes on to feature the struggle between a greedy energy company determined to suck the world dry and a ragtag group of ecoterrorists symbolically represented by Aerith and her gutter flowers.

Another way to reveal the world is through flashback or backstory. In *Uncharted 4: A Thief's End*, flashbacks communicate valuable information to deepen our investment in the characters and story line. After starting the game during a boat chase scene in media res, Nathan Drake falls overboard and blacks out. We have our first flashback: Nathan sitting on a cot in an orphanage after getting into a fight. We learn more about his background. He's a talented kid who resents being there and who looks up to his older brother who has a reputation as a troublemaker. The scene expands what we know of the characters and complicates their goals. They are brothers. They had a tough upbringing at an orphanage that assumed the worst of them. The treasure that they're searching for has a history back to their jilted childhood. We want to know what happened to them. The player is more eager to help them on their quest and to get the two somehow reunited.

But communicating information about the world through exposition and backstory poses some risks. It can disrupt game flow and draw attention to itself. Bob McKee's *Story* argues that flashbacks need to follow the same rules of introducing exposition as the present story: "A flashback can work wonders if we follow the fine principles of conventional exposition" (341). The key, McKee argues, is to dramatize the exposition. If the exposition is long, dull, unmotivated dialogue solely for the audience's benefit, it's going to be dull anywhere—especially in a flashback. Instead, create a flashback with its own plotted dramatic structure. In short, a scene worth diverting our attention that is dramatic as it reveals expositional information. If it doesn't do this, the audience will feel that the flashback is useless, uninteresting. If a video game story line is linear enough and the mechanics don't invite diversions or world exploration, dull cutscenes in flashbacks can similarly disrupt the player experience, prompting the player to skip if the information about the world is not important enough to key game mechanics. If you're going to have diversions to deliver information, it has to be dramatic and significant enough to be worth the risk.

One solution to presenting expositional information is to make the past playable. That way the flashback isn't a diversion interrupting the gameplay but a more integral part of the game itself. One of the reasons that the expositional information in *Uncharted 4: A Thief's End* doesn't seem superfluous is because we play through it; it's not simply a cutscene but a playable chapter with its own structure and goals. Interactivity and playability increase the player's investment in expositional information, making the player less likely to skip it.

Video games can incorporate other media to deliver information about the world without relying on forcing exposition into the overall narrative. We can learn from other artifacts, from objects or diaries or otherwise interacting with the world. *Her Story* communicates exposition through recreating a 1990s-era computer interface with a database of searchable interrogation videos. Because the interface is designed to search, the player learns exposition in a way that feels natural. Dotnod's *Life Is Strange* plops us into a familiar enough school environment, but we learn the particulars (the social hierarchies in the art school, the Pacific northwest setting, the death of a local teenager) through the character's interaction with her environment. We read her journal and notebooks, we interact with posters in the classroom, and we bump into other students. When Max starts manipulating time, we're literally invited to expand what we know about the world again and again till a full tableau of the setting helps us understand what we need to do next to solve the game's mysteries.

Large-Scale Fictional Worlds/ Collaborative Worldbuilding

Much of our discussion about setting and worldbuilding thus far has assumed an individual writer and not a writer working collaboratively or with a group of other writers. In larger games, the task of worldbuilding is often a group enterprise, where the writer needs to use the tools that we've outlined here and apply them to a collaborative environment. That means sometimes relinquishing control or finding a particular niche of the worldbuilding to contribute to. Most of our creative writing classroom experiences will have a collaborative element to them, making skills like flexibility, organization, adaptability, and collegiality just as important as your skills as a writer. These kinds of collaborative environments will better prepare you for writing in the workplace, whether that be on an indie game with a handful of writers to a large AAA studio or any other multimodal writing environment.

Some of our suggestions for communicating the world, for example, change if you're working in a collaborative environment. In a large studio, writers may be responsible for some of the work communicating the setting, but most of the time this will be left to a designer or artist. Much like Hollywood, the writer provides a framework of the world that leaves itself open for collaboration. If you're describing an opening cutscene, for example, don't make suggestions about the way that the scene should be shot but instead describe concretely the particulars of the scene that the designers need to be aware of. This way, the designers and animators are free to imagine the opening scenes however they want to bring the world alive visually for the gamer. Be careful about providing any direction on how the game should look. If you're working on film-like cutscenes, a writer intruding on that space is like a micromanager telling visual artists what to do. You may have an idea how things might look but you should communicate it only if invited.

The work of creating or imagining the world may also be collaborative. An excellent resource for collaborative worldbuilding is Trent Hergenrader's *Collaborative Worldbuilding for Writers and Gamers*, a comprehensive textbook about the process. Hergenrader guides the writer through the various stages of worldbuilding, starting with brainstorming, developing the world's foundation and framework, identifying the world's structures and substructures, writing a metanarrative, and finally producing a catalog for reference.

Although Hergenrader's book is primarily for developing fictional worlds, the principles that he outlines are equally applicable to real-world locations. A game set in Chicago, for example, could just as easily include a historical metanarrative and research about the actual structures present in the city. A catalog from which other writers can draw from would be just as essential as for a fictional location. Because even though information about Chicago may be widely available, it helps to have a document that everyone is working from that can serve as a touchpoint for narrative development to help keep writers on track.

The principles of collaborative worldbuilding can also be helpful for someone working individually on their own project. Some of Hegenrader's examples draw from worlds that were individually authored, such as George R. R. Martin's world of Westeros or Tolkien's Middle Earth. Maps, structures, a metanarrative: all these are equally important for a game developer working on their own to help organize and breathe life into their world.

Overview

In this chapter, we've introduced how to activate setting, how to research to find idiosyncratic details, and how to communicate fantastical or realistic worlds effectively to an audience. We have also introduced some basics about worldbuilding to use in collaborative or individual environments. The next chapter will address how to write compelling dialogue to flesh out the characters that will inhabit your worlds.

6

Dialogue

A father, home from work, late, still on his cell phone about a botched construction job.

"Tommy, Tommy. Listen to me. He is the contractor, OK?"

A clock ticks in the background. His daughter, his only daughter, only family of any kind, lies on the couch, hands tucked under her short hair.

"I can't lose this job."

Silence. He flicks a switch. His daughter stretches one arm, then another.

"I understand. We'll talk about it in the morning."

He tosses the cell phone on the coffee table and his daughter yawns and opens her eyes. He must look terrible: eyes shot, worry-worn, his faded t-shirt and jeans remnants of better times.

"Scoot," he says.

"Fun day at work?"

"What are you still doing up? It's late."

"Oh crud. What time is it?"

"Way past your bedtime."

"But it's still today."

She crouches down and retrieves a square clamshell box hidden under the couch. It's small and gray, half the size of a shoe box.

"Honey, please not right now. I don't have the energy for this."

"Here."

"What's this?"

"Your birthday?"

The man takes the box in his hands and slowly opens the lid.

"You kept complaining about your broken watch. So, I figured, you know."

He removes the watch and straps it to his arm.

"You like it?"

"It's nice, but." He taps the watch face and brings it to his ear.

"What?"

"I think it's stuck. It's not—"

She lunges for the watch. "What? No, no, no. Oh, ha ha."

He sits back. Dad joke accomplished. "Where'd you get the money for this?"

"Drugs. I sell hard core drugs."

"Oh good. You can start helping out with the mortgage then."

"Yeah, you wish."

He turns on the television. Soon, his daughter is asleep. He cradles her in his arms, her rumpled band shirt and flannel pajamas soft against his arms. He walks her to her room and lays her in bed. He tucks a lick of hair behind her ear. One last look before he turns the light out.

This opening scene from *The Last of Us* introduces us to Sarah and Joel, a father-daughter family who are caught at the beginning of a zombie-like pandemic. Initially, Sarah is the playable character. We navigate through her house the next morning, trying in vain to find her father. When he finally bursts through the backdoor, the pandemic is in full swing. The neighbors are infected. They must find her Uncle Tommy and battle their way through to safety. But safety never comes.

Why start the game with this tender scene on Joel's birthday? The dialogue establishes character, showing their relationship. Their short back and forth when he pretends that the watch isn't working reveals both that his daughter cares about him and that they can joke with each other. Where did she get the money? Selling "hard core drugs." The subtext is that she's found a way, that she has enough autonomy to buy him a present, this father who works late and who is trying to support her. "Oh good. You can start helping out with the mortgage then." A reversal. Far from being shocked, the father knows his daughter and trusts her. The teasing implies almost the opposite of his words: instead of wanting her to help pay the bills, he would do anything in his power to protect her.

When the zombies come, our sympathy has deepened. We identify with these characters and care for them. The dialogue heightens the emotional impact of the introduction and strengthens our investment in Joel. We better understand his motivations and his willingness to put his life on the line for Ellie, a young girl near his daughter's age, later on in the game.

In fiction, dialogue is one of many tools of direct characterization. A writer can choose to describe physical characteristics, to put the character in action, reveal the character's thoughts, or have the character open their mouth to speak. In a visual medium such as film or video games, dialogue becomes even more important since it's the primary way we come to know a character. Dialogue, if used correctly, can give us the best sense of who our characters are.

So, what makes good dialogue? We all know the frustration of playing a game or watching a movie and hearing the clunk of bad dialogue. It sounds unnatural or forced. Maybe it's overly sentimental or cliché? Sometimes the dialogue isn't economical enough or redundant, explaining what we've already seen on screen. Even some iconic films suffer from this. A frequent complaint from actors in the early *Star Wars* series was George Lucas's leaden dialogue. Pleading from Harrison Ford and Carrie Fisher led to some changes but other clunkers still slipped through: "But I was going to Tosche Station to pick up some power converters!" Luke says. Delivered in a high whiny voice, this line is often mocked as one of the worst in *A New Hope*. "I have placed information vital to the survival of the Rebellion into the memory systems of this R2 unit," Princess Leia says. Clunk. How did the actors even deliver these lines with a straight face?

So, dialogue is important. It can convey mood, conflict, personality, state of mind. It is NOT simply replicated speech. Good dialogue is selective and memorable, a little like poetry.

Despite the many legitimate criticisms of Ernest Hemingway and his life and work, he introduced a concept in his nonfiction book *Death in the Afternoon* that for better or worse has come to dominate contemporary short fiction:

> If a writer of prose knows enough about what he is writing about, he may omit things that he knows and the reader, if the writer is writing truly enough, will have a feeling of those things as strongly as though the writer had stated them. The dignity of movement of an iceberg is due to only one-eighth of it being above water.
>
> (Hemingway, *Death in the Afternoon* [1932])

This "iceberg theory" has caused some readers to review Hemingway's work, looking for subtext, such as the "simple operation" in his infamous short story "Hills Like White Elephants." The metaphor of the iceberg has many limitations but it's useful for understanding how the job of the writer is to convey more than just surface information. The *real* story is what lies beneath. In "Hills Like White Elephants," the two conversing lovers talk benignly about the weather, the coming train, what they'd like to drink, till the unnamed narrator says, "It's really an awfully simple operation, Jig . . . It's not really an operation at all." This bit of dialogue reveals that their conversation may be about something more than just passing the time before a coming train. The woman is considering an abortion. There are stakes here, and for a traveling expat American couple in 1920s Europe, those stakes can be enormous. The iceberg looms large.

One of the cardinal rules in writing good dialogue is to avoid forcing exposition. The reasons for this are many but the most compelling is that it sounds unnatural, like the speaker is for whatever reason giving an information dump. Ears that are attuned to the natural cadences of speech will recognize its awkwardness. Most fiction by the time it's published has eliminated exposition in dialogue although in visual mediums like film or video games, it's more commonplace. In fiction, the author can move exposition to a character's thoughts or to description. But in video games, there's no easy way to do this. Sometimes the temptation to force exposition into dialogue can be too great. The writer needs to reveal information that she wants the audience to know and they deem it worth the risk.

So how can expositional dialogue feel natural to the player? Well, a player's tolerance for exposition grows if the source is from one that they would expect to deliver it. For example, if a character is a medical doctor giving a diagnosis or a scientist explaining the origins of a new superbug or a technician telling us how a time machine works, we're more willing to go along for the ride. In the film *Back to the Future*, Doc explains several times the full plot of the film, sometimes with charts and timelines as reference for Marty's (and our) benefit. If the main protagonist is in the dark and wants to know something, the character that they turn to can often convey expositional information in a way that feels natural. Consider the following passage from a collaborative Twine game written in one of our classes. The game is called *Charted*, about a near-future AI service provided for students trying to decide on a career. The player arrives at the high school guidance counselor's office ready to make some decisions. The guidance counselor explains:

I'm here to help you decide what to do with your life. Charted is a wonderful new service but it can be expensive, depending on what pathway you choose. Some of the variables will also influence the price, largely because of the complexity of the AI predictability model. Relationships—now relationships are the worst. I'm mostly here for your career path, but since partners can have a tangible influence on your future choices, we also offer that option for our graduating seniors. But you'll have to get their consent if you want that pathway package.

Because the students have used a guidance counselor as the character, her dialogue can effectively communicate exposition without it sounding forced. It's her job to inform the player about the various options. We now also know the rudiments of the game's plot: the player will have to choose one of several pathways that will chart the player's future. RPGs and other quest-based video games often use a similar tack when introducing plot elements. The player encounters an NPC or experiences a cutscene near the beginning of the game that lays out what the player hopes to accomplish. The game then becomes less about figuring out what is going to happen—the player needs to repair the Elden Ring (*Elden Ring*) or stop the rogue Spectre Saren Arterius (*Mass Effect*) or repair time to prevent the future destruction of the world (*Chrono Trigger*)—and more about the journey and how the player may realize her goals.

Video games also have other reasons to convey exposition in dialogue. Because video games have the added task of communicating information about game mechanics or intermediate goals, avoiding exposition in dialogue is rarely an option. We need to know how to jump, shoot, gather health, mine, or heal. We need to know where to go, how to solve puzzles, or what the next steps in a game might be. Again, this is best done more naturally, through a character whose role mentors the playable character. In *Horizon: Zero Dawn*, Aloy learns not only about her identity and basic information about the world from her mentor and father figure Rost, she also learns how to hunt and survive in an introductory tutorial that's fully integrated into the story line. In *BioShock*, almost everything we learn about the underwater world of Rapture is through Atlas, a voice who speaks to the PC Ryan via a portable shortwave radio. Atlas asks us to trust him, teaches us how to use plasmids, instructs us about some of the basics of the world (little sisters, big daddies, splicers, etc.), and through dialogue, helps keep us alive. It feels natural to the gameplay and we never question his omniscient counsel as he guides us through one scrape after another. This is why, near the end of the game, we are so shocked to find out his identity, revealing how much Atlas's dialogue has manipulated us all along.

Overuse of dialogue for these kinds of tasks in narrative video games can feel forced, so finding ways to convey information naturally is key to a seamless gaming experience. If instructions or exposition in dialogue are too obvious, it will take the player out of the game or even open up the game to criticism or ridicule. More players will be likely to skip sections, sometimes leaving the player without essential information for completing the game. In short, make sure that conveying instructional information or exposition in dialogue is absolutely necessary and that the stakes are high enough in the gameplay to make it worth it for the player.

Dialogue and Plot

Good dialogue can not just reveal the plot; it can also move the plot of a narrative game forward. In dialogue-heavy games especially, plot is tied to the conversations between characters. Dialogue is often an expression of character desire; when a character enters into a conversation, those desires come into contact with opposing desires, resulting in tension that drives the plot. When characters disagree, they create obstacles for each other, obstacles that are often overcome through more dialogue or a change in behavior in the game.

Take, for example, the dialogue-heavy indie game, *Doki Doki Literature Club* (*DDLC*). Designed to mimic anime-inspired visual romance novels, *DDLC* deconstructs the conventions of its genre through the dialogue. The four female NPCs have dialogue specific to them that becomes more pronounced and exaggerated as the game progresses. At the very beginning, Sayori invites the player to join a new club. They are childhood friends who are now in high school. Before the player walks Sayori to school, she asks, "Are you proud of me?" to which the player replies, "For what?" "For waking up on time!" Sayori says. Her dialogue is bubbly and innocuous. She's finally getting up early to go to school. So what? But to her it's important. "You never even said anything about it," she says. She badgers the player till they say, "I'm proud of you, Sayori."

This first dialogue sets the plot in motion by both revealing Sayori's character and setting up the first stakes of the game, whether or not the player will join the literature club along with three other girls: Yuri, Natsuki, and Monika. All three have dialogue specific to them: Yuri is preternaturally shy, Natsuki is outspoken and critical, Monika is bossy. As the game progresses, the dialogue becomes more exaggerated and sometimes flirtatious. Sayori's difficulty waking up in the morning turns out to be symptomatic of depression. Yuri's shyness becomes obsessive, Natsuki's assertiveness becomes snarky and vulgar, while Monika's bossiness becomes dictatorial. The personalities clash at the literature club, with the girls vying for the player's attention. Sayori relapses into depression the closer the player gets to her. "I like you so much that I want to die!" she says. When one day she doesn't show up to walk to school, the player goes next door and finds she has hung herself.

To move the plot forward, dialogue needs to express character desire and introduce conflict. A statement like "I like you so much I want to die!" forces the player into action leading up to the climax. They don't want Sayori to die, so they will make decisions reacting to her dialogue. Do they hug her and confess their love for her? Or would it be better for her mental health to remain friends? The dialogue pushes the action forward. Unfortunately in *DDLC*, there is nothing that the player can do. Depression is a sickness that the player cannot cure. Whatever they choose, Sayori will die.

Voice

One of the most important skills a writer needs to develop is writing voice in dialogue. Cultivating voice requires listening for colloquial nuances to capture the essence of a character succinctly and economically. Distinctions in character should be apparent through the language that they use in a game. If all the characters end up sounding the same, the writer hasn't done a good enough job differentiating character voice.

The best way to ensure that characters sound unique is to use language that is specific to them. Remember, character is made up of words, and those words can change what we know and understand about a character. That's why character sheets, especially in longer narrative-driven games, can be so essential. While any good game should allow for characters to be dynamic and change, that rarely means that their way of expressing themselves through dialogue will change. If you're writing a character who has difficulty expressing their emotions and communicates in fragments with a simple lexicon of single-syllable words, we're going to feel violated at the loss of character continuity should they suddenly start in on a flowery Shakespearean monologue. Keep in mind who your characters are and have a reference sheet if you're using multiple writers to keep the voice consistent and distinct.

Research is one of the best tools to improve a character's voice. Say your game has a character who is an equestrian. Now if you don't know anything about being an equestrian, you could just rely on your own basic knowledge: "I need to go practice my riding." But you'd miss an opportunity to use discipline-specific language that could make that character stand out. Just a quick Wikipedia search will give you a wealth of language that you can use in dialogue to effectively differentiate one character from another. Harness, lariat, shank, curb bit, dressage—do research on the character's job and suddenly you have a lexicon unique to the character: "I need to practice my canter" or better yet, something with a little more personality: "Lief says I'm not putting enough emphasis on the first beat of my canter. Gotta get my horse more uphill, stronger on the back hind leg." Be careful when using it, though; sloppy usage of discipline-specific language will sound disingenuous. But if you come from a certain background (say, equestrianism) you'll likely be attuned to the nuances of the characters' speech and employ discipline-specific language naturally. Finding speakers who represent a particular group can be essential to ensuring authenticity. This becomes especially necessary in areas of race, gender, or with any group that has strong identifying characteristics (particular religions, countries, etc.). There's nothing more violating than hearing language from someone sharing your own subject position misrepresented in a game through cliché or stereotype.

That's one of the reasons you'll want to avoid writing in dialect. It can be offensive and demeaning. There are solid historical and cultural reasons for this. During the realist era in the late nineteenth century, dialect was used to distinguish characters of a certain race. White authors would often (mis) represent African American characters through dialect. When we first hear Jim speak in Mark Twain's *Huckleberry Finn*, for example, he says, "Who dah?" for "Who's there?" He uses language like "gwyne" for "going to" or "sumf'n" for "something." These words today sound painfully stereotypical. But at the time, Twain claimed his use of dialect was realistic and accurate. For the reading public in the predominantly white North in the mid-to-late twentieth century, Twain's depiction of Jim became a devastating representation of African Americans. Having never heard Black idiom before, they bought it as the way that Black Southerners truly spoke. The convention of representing Black characters this way became so pervasive that even Black Northern writers who never spoke in dialect felt compelled to use it when representing their characters. The implications of dialect today have become so problematic that it has caused some contemporary revisions of Twain's work, removing the dialect and instead relying on lexicon, syntax, and context to communicate Jim's character.

A contemporary solution to convey significant detail about a character's language is to provide information in a tagline or thoughts. In Octavia Butler's novel *Kindred*, the protagonist Dana starts inexplicably traveling through time to visit one of her ancestors, Rufus: "'You lay a hand on me, and I'll tell my daddy!' His accent was unmistakably Southern, and before I could shut out the thought, I began wondering whether or not I might be somewhere in the South. Somewhere two or three thousand miles from home" (21). Instead of writing in Southern dialect, Butler conveys location by commenting on his "unmistakably Southern" accent and by Rufus's threat. From this point on, we almost subconsciously read Rufus's dialogue with a Southern accent even though it's written in pretty much standard English.

Another reason to avoid dialect: if you're eventually going to be working with voice actors, it's up to them how they will want to employ accent or riff for vocal nuance. Putting dialect in speaking parts is like insisting that your words be said a certain way, often forcing unnatural responses from the actors. It takes away their autonomy, their ability to imagine themselves in a character. But add interesting lexicon and syntax that sound specific to a certain character, and they're likely to expand on the role and make it all that much better. This is why screenwriting textbooks counsel writers to avoid dialect: it makes assumptions about how the character speaks that the director and actors may not share. For example, in *Mafia*, developed in 2002 by Illusion Software, we begin with protagonist Tommy Angelo being convinced by wise guys Paulie and Sam to help them out of a jam. Their environment is clearly a fictionalized Manhattan, and the boys talk in old-fashioned slang as cars meant to evoke the 1930s whiz by. It's not stereotypical accents that ground us in time and place but the accumulation of significant details—combined with a gritty lexicon seemingly lifted from *The Godfather*—that result in a rich opening scene.

A word on diversity: sometimes our advice to students to diversify their characters leads them to arbitrarily choose characters far from their own subject position. Often this will result in stock characters using cliché language in ways that can be harmful or problematic. This isn't what we mean by diversify. Instead, think within the world and character framework for your game. If you're using a large-scale fictional universe like *Star Wars*, this may give you some latitude to use diverse characters but be careful; don't let aliens be stand-ins for ethnic stereotypes. If the characters start sounding like Jar Jar Binks—who even in 1999 came under fire for sounding too similar to a stereotypical minstrel character—then you have a problem. Think carefully about how you want your characters to speak in whatever universe they inhabit. Try to imagine diversity in how characters speak linguistically; maybe you have a character who speaks in fragments. Another who talks too much and uses a lot of profanity. Or a character who uses primarily logic to function and who doesn't understand subtext. Contrast that with someone who is sarcastic and rarely means what he says and you have character gold, like *Star Trek's* Spock and McCoy, an onscreen rivalry that kept fans coming back again and again to the franchise.

And the absolute *best* way to ensure diversity of characters is to have a diversity of writers. If you're working in a writing room, look around. If everyone's like you, you're going to be more likely to have a problem when it comes to finding relatable characters from outside your collective subject positions. Diversify your writing room and you diversify how your characters will sound.

Conventions for Writing Better Dialogue

1. Be selective.

Remember, dialogue is not just replicated speech. Even radio programs will carefully edit dialogue to its bare essentials, getting rid of filler words or pleasantries that dilute the listening experience. In a visual medium like games, excess dialogue is rarely necessary. Avoid long soliloquies or unnecessary speeches. "You sly dog, you caught me monologuing!" the supervillain Syndrome says during *The Incredibles* after an overlong speech. Good dialogue is about compression.

Consider the following scene:

"So did you have a good day at work today?" Mike said.

"It was OK, I guess. I mean, yeah, you know, I had a lot of work that I didn't get done last night because I was out with you but other than that, no problem."

"Hmmm, about that. You know you're my girl and everything but you probably want to just keep what we're doing on the down low. You didn't say anything to anyone did you?"

"No, honey of course not. So, what do you want to do today?

"I don't know, maybe drive around town a little. That is, unless you got something else better to do?"

This dialogue starts with an unnecessary pleasantry. Generally, you want to avoid dialogue that feels like the characters are spinning their wheels without anything better to do. Don't talk about the weather, what you did at work, ask how was your day, etc. etc, unless it's absolutely essential to the story or game. It's not that people don't say these things; it's that they say them every day, over and over again, and rarely do the conversations imply anything larger than themselves. Already from the outset, the author has chosen a point of entry that is banal and mundane. Start the conversation elsewhere when something's more at stake.

The diction here is also full of filler words. "So," or "It was OK," "I mean, yeah, you know" don't do anything but cushion the dialogue. Again, people do talk like this and it's even possible that a voice actor may reintroduce filler words when they're adapting the dialogue to their character. But avoid writing them because they, you know, don't accomplish much.

The question-and-answer format is also problematic, resulting in dialogue that feels forced. People sometimes ask each other questions but rarely do they answer directly. Q+A works for an interrogation but even then you'd get more tension out of an interrogation if one of the parties wasn't cooperating. Lobbying questions back and forth feels like a writer spinning their wheels. They don't know how to get the characters to interact so they force dialogue through banal questions and answers.

But the most problematic element of the exchange is the forced exposition that we get in the second paragraph. The girl introduces information through dialogue that's important to the scene but it's there for the reader/players' benefit and not for the characters. Why does she reveal that she was out with Mike late last night? He would know that she didn't do her homework. It's also obvious that she's his girl. The writer is slipping in the information solely for the reader/player.

2. Avoid sentimentality and exaggeration.

Like any writing, dialogue must earn its emotion. Nothing irks more in a scene than overly sappy declarations of love or emotional outbursts that seem to come out of nowhere. Like Hollywood popcorn movies, many AAA video games suffer from this. The game comes to an emotional moment, say a significant death, and the writer feels an obligation to emphasize its importance by overwriting the dialogue. In these situations, it's best to exercise restraint. If you've done a good enough job establishing the character, the player will feel the emotion of that character's loss (and not just death—it could be a departure, moving on, or a change in status like a break-up). Sometimes a significant loss requires no dialogue at all. In the 2013 adventure game *Brothers: A Tale of Two Sons*, Starbreeze Studios chooses to convey the older brother's death through the younger brother's attempt to revive him with water from the Tree of Life. When he realizes that he is already gone, we get only one word of dialogue: the brother's name: "Naia." Because of gameplay that emphasizes the brothers' strong bond—you play each brother using one of two thumbsticks on the same controller—the minimalist dialogue is all we need. We feel Naia's loss every time we reach for the older brother's thumbstick and feel his absence. And even if we *did* have more dialogue, it would've been in the unintelligible language that Starbreeze Studios created for the game. The context and gameplay are enough for the player to understand the characters as they navigate this world to save their ailing father.

Writer Mimi Schwartz calls this "cool language." When a scene has high emotional stakes, the writer needs to cool down the language for the greatest emotional impact. The hotter the emotion, the cooler the language.

3. Do more than one thing at once.

Truly effective dialogue usually has multiple functions. A single line of dialogue can convey instructions, contain subtext, move the action forward, and characterize. Dialogue-heavy games rely on dialogue to provide texture and direction to the gameplay. Night School Studio's 2017 supernatural walking simulator *Oxenfree*, for example, uses dialogue to define relationships, advance the plot, and reveal character. A simple 2D third-person game with simple graphics, *Oxenfree* uses dialogue in speech bubbles above characters' heads like in graphic novels. The playable character, Alex, always has two or three bubbles above her head and the player advances the game by choosing one of them. Near the beginning of the game, her new stepbrother Jonas joins Alex and their mutual friend Ren on a trip to an island to meet up with some other friends for the weekend. Their parents have recently married and Jonas and his father have moved into Alex's house. They don't really know each other yet and the first few interactions are punctuated by awkwardness. When they arrive at the island, Jonas asks Ren to leave them so Jonas and Alex can talk.

Jonas: "Can I have two real quick minutes with Alex for a second? And you can run up and meet your friends?"

Ren: "Uh, really?"

Then three speech bubbles appear above Alex's head. You can choose "What's wrong with Ren being here?," "Is something wrong?," or "Don't be weird already Jonas." Each one of the choices will have a different effect on Jonas, altering the game. Choose to be rude and snarky, "Don't be weird already Jonas," and you will increase the tension, making your newly blended family all the more awkward. Choose to be nice and an image of Alex's head will appear in a bubble above Jonas, showing that you're on his mind and becoming closer. Here, the dialogue accomplishes multiple things at once, providing direction in the gameplay, characterizing both Alex and Jonas, and defining their relationship. You can end the game as close siblings or as rival ones. The dialogue also often conveys subtext. Jonas's initial request creates tension since he asks Ren if he can have a couple minutes, not Alex, as if he is more concerned about Ren's friendship than he is hers. The "two real quick minutes" suggests that it's not much of an imposition, but it also implies that there's a significant reason he would like to talk to her, something that perhaps requires both privacy and secrecy.

4.   Avoid unnecessary repetition and clichés.

Nothing kills dialogue like cliché or hackneyed language. You can have the most interesting backstory and motivation in the world but if the dialogue is wooden and predictable the character will seem insipid and shallow. Sometimes we will have students who will argue, well, I'm trying to *show* that my character is shallow by *using* clichés. Or they'll pull the "I-know-somebody-like-that" trump card. And it's true that in real-life people sometimes talk in clichés. But they are usually so much more interesting if they don't. The more unique the dialogue, the more unique the characters. A shallow character with unique dialogue is more interesting than a complex character who speaks in clichés.

A similar but related problem is repetitive dialogue. This can be a challenge in games where gameplay is repetitive or recursive. It's common enough that some indie games draw attention to how unnatural some repetitions truly are. In the game *Undertale*, for example, a shopkeeper refuses to repeat a dialogue option if you've already selected it. This is particularly funny because of how common it is in games to be able to replay the same dialogue over and over as if it had never happened before, even when the game tries to be realistic.

Repetition becomes especially difficult to avoid in situationally triggered contexts. For example, if you have a conflict/fight or you pass an NPC, those actions will trigger dialogue, often called "barks." In some situations, actions will trigger random barks, requiring the writer to write a number of different lines of dialogue for the same action. For example, if your character bumps into someone accidentally, you might hear, "Watch where you're going?" or "Excuse me" or simply "Hey!" Some barks are to inform players and their frequent repetition in a game can lodge them in your brain. An omniscient "Green warrior needs food badly!" echoes every time a warrior is low on food in Atari's 1985 dungeon crawler *Gauntlet*. Contemporary games that seek to present a seamless, more realistic world and gaming experience need to vary the barks and avoid repetition. These situationally triggered barks will often do more than just inform the player. They can characterize NPCs or develop the texture of the world. They can introduce a situation to try to get the PC to engage in dialogue. They can also provide humor: "My hotel's as clean as an Elven arse!" the innkeeper says in Bioware's *Baldur's Gate*. But humor can wear thin if overly repeated. A joke's usually funny only once.

Overview

In this chapter, we've shown how dialogue functions in multiple game contexts, how a writer can use dialogue to reveal character, advance the plot, or convey subtext. Learning to write good dialogue requires patience and observation. You have to become attuned to the cadences of speech and learn to recognize when people are saying multiple things at once. The next section will first explore the variety of video game genres and then look at elements of game design where you can apply the creative principles of character, plot, worldbuilding, and dialogue to designing your own games.

Part II

Game Design for Creative Writing

7

Demystifying Genres

Unpacking Genre

As artists who want to write video games, what must we learn in order to begin? Although the truncated history of video games discussed in Chapter 2 and the creative techniques explored in Chapters 3 through 7 allow us to imagine both this medium's origins and the vast storytelling potential, writers would be remiss not to explore what video games have accomplished historically and what they still might accomplish in the future. To that end, it's essential to survey a diverse array of video game genres and forms. In the following sections, we'll unpack the aesthetic language of key video game genres while providing multiple examples of games to play and study.

Role-Playing Games

When it comes to narrative in games, no genre embraced story as early as the role-playing game. Developed along parallel tracks in the West and Japan, this genre evolves from *Dungeons and Dragons* tabletop role-playing. The focus here is less on button mashing or quick reflexes and instead on strategy and reading pages upon pages of text—or more recently watching hours of cutscenes. These games traditionally take place in medieval fantasy settings—interesting examples include *Skyrim*, *Final Fantasy IX*, or *Tales of Symphonia*—but there's also a rich history of RPGs that borrow heavily from sci-fi—including *Final Fantasy VII*, *Star Wars: Knights of the Old Republic*, and *Mass Effect*. Some RPGs even focus on mirror house images of our own society—*Earthbound*, *Yakuza: Like a Dragon*, or *Disco Elysium*—while others borrow from all of the above—*Chrono Trigger* for example.

The first significant Japanese role-playing game to both sell well and earn critical acclaim is 1986's *Dragon Quest*, released on the Nintendo Entertainment System. Cocreated by the trio of Yuji Horii, Koichi Sugiyama, and Akira Toriyama—the famed artist responsible for *DragonBall Z*—the original *Dragon Quest* is a top-down 2D game in which the player guides a knight in search of a damseled princess. It's light in terms of story compared to later Japanese role-playing games (JRPG), but it was revolutionary at the time to wander through villages while speaking to people at your leisure, reading their dialogue in text boxes reminiscent of a comic book. While exploring the wilderness, the screen will randomly cut to a static shot of an enemy and an encounter ensues. Instead of steering your avatar head first into battle like in *Super Mario Bros.*, you instead select their actions from a menu of options. If victorious, the player earns experience points. Gain enough of these, and the player levels up from one to two to three and so on. With each new level, your attacks grow stronger, as does your defense. This template is the foundation of nearly every JRPG—even today as more and more trade the menu-based commands for action-oriented control. The original *Dragon Quest* sold two million copies worldwide and spawned ten sequels in addition to multiple spin-offs. It directly led to the Final Fantasy series launched the following year, and its influence can be felt in both 1990s JRPGs like *Breath of Fire* and *Chrono Cross* and more modern titles like *Xenoblade Chronicles* and *Bravely Default* or even Western games that adopt the JRPG-style like *Pier Solar* and *Crosscode*.

Western RPGs, on the other hand, thrived in the home computer market of the 1970s and 1980s. Instead of taking their cues from anime and manga, RPGs in the West typically employ a style more suited to a writer like J. R. R. Tolkien. Text-only games like *Zork* and *Colossal Cave Adventure* provide proto examples of the genre, but most pinpoint 1981's *Ultima*, *Wizardry*, and 1984's *King's Quest* as the period when these games entered the mainstream. Early Western RPGs resemble their Japanese counterparts, but they often focus more on battles, stats, and first-person mazes more than the soap opera theatrics of JRPGs.

In terms of storytelling tropes, there are a few constants across most RPGs. Most of these games focus on large, interconnected casts and interpersonal relationships. Likewise, there's almost always an epic story spinning out over multiple acts and sometimes multiple time periods in the vein of *Game of Thrones*, *Great Expectations*, or the original *Mobile Suit Gundam*. Narrative typically comes in the form of the player chatting with townsfolk or during scripted cutscenes similar to film. The chorus-verse-chorus of most RPGs is explore a town, fight through a dungeon culminating in a boss battle, watch a cutscene, repeat. Many of the best RPGs explore complicated topics like philosophy (*Xenogears*), PTSD (*Final Fantasy VII*), and pacifism (*Undertale).* Writers who yearn for an immense canvas with a Dickensian number of characters should gravitate to the RPG, especially writers with a soft spot for fantasy in the vein of Nnedi Okorafor or sci-fi like Octavia Butler.

Action/Adventure

If you consider *Spacewar!* the first video game, then action is the medium's first genre. Adventure games can tell simple stories like *Super Mario Bros.* to motivate the player and scaffold the action—the squat Toad telling Mario, "Sorry, but our Princess is in another castle"—or they could interrupt the gameplay with hours upon hours of cutscenes mimicking big-budget films. *Metal Gear Solid IV*, for example, has a continuous sequence of cutscenes that runs for seventy-one straight minutes! But it's useful to think of the action/adventure genre in terms of action films. The canvas is extremely wide, supporting both something straightforward like *Raiders of the Lost Ark* and something more complicated like *Everything Everywhere All at Once*.

Although the 1980s and early 1990s featured action/adventure games with crudely ambitious narratives—*The Legend of Zelda* and *Illusion of Gaia* come to mind—those games tended to favor RPG elements, and it wasn't until 3D graphics became ubiquitous that action/adventure games began to embrace narrative at the level of their RPG counterparts. One noteworthy example is 1998's *Metal Gear Solid*, a PlayStation game directed by Hideo Kojima, one of the first developers elevated by fans and critics to the status of auteur not because of their attention to gameplay like Nintendo's Shigeru Miyamoto but because of Kojima's experimental brand of storytelling. *Metal Gear Solid* tells the story of Solid Snake, a retired spy recruited for one last mission who must infiltrate the Shadow Moses military base taken over by American soldiers turned terrorists. In many ways, the original *Metal Gear Solid* is derivative of a dozen action movie blockbusters—*Escape from New York* and its protagonist, Snake Plissken, chief among them. But compared to most console action games from the late 1990s like *Crash Bandicoot*, *Spyro the Dragon*, and *Super Mario 64*, *Metal Gear Solid* stands alone as wildly ambitious. Kojima's breakthrough is peppered with long cutscenes focusing on the need for nuclear disarmament, the horrifying legacy of Hiroshima and Nagasaki, and the dangers of genetic tampering. At the same time, there are multiple meta moments that poke fun at its self-serious narrative, like when Snake confronts a psychic and the player must physically unplug her controller from the PlayStation to break the villain's mind-link. It's a subversive gonzo tale in the candy coating of a Hollywood blockbuster, and it's easy to draw parallels not only to the next two decades of Kojima games but also to the entire generation of meta-action games that followed—including *No More Heroes*, *Eternal Darkness*, and *Spec Ops: The Line*.

Much like the RPG, narrative in action/adventure games evolved in parallel lines—one developed in Japan for consoles, while the other developed in the West on PCs. Released two years after *Metal Gear Solid*, *Deus Ex* launched on the PC and was directed by Warren Spector of Ion Storm. Upon first glance, *Deus Ex* might resemble a simple shooting gallery like *Wolfenstein 3D* or *Quake*, but this first-person shooter spends a surprising amount of time on its atmospheric cyberpunk narrative that borrows liberally from *The Matrix*. *Deus Ex* reflects on class inequality, limited pandemic resources, and even the dangers of artificial intelligence. Although *Metal Gear Solid* features two different endings based on whether or not the player surrenders to torture, *Deus Ex*'s endings are triggered by the player's narrative choices. At the game's end, players can opt to bind themselves to an AI entity, plunge society into a new dark age, or cover up everything they've learned while simultaneously handing over the world to the Illuminati. While RPGs often take inspiration from fantasy novels or manga and anime, action/adventure games often emulate films—especially the work of Ridley Scott and John Carpenter—or sci-fi novels like William Gibson's *Neuromancer*. *Deus Ex* remains a landmark title to this day, and we can see its DNA in more modern first-person shooters like *BioShock*.

Tropes are difficult to narrow down in a genre as amorphous as action/adventure, but generally these games interrupt their action with bursts of narrative that usually come in the form of cutscenes—when control is stripped from the player and they watch the action unfold like a movie—or text boxes like in a comic book. Occasionally, narrative action games will forego cutscenes and text altogether, allowing the player to move freely while other characters speak and progress the story—see 1998's *Half-Life*. A 3D game like *Uncharted* interrupts its gameplay for frame narrative cutscenes that mimic the tone of Indiana Jones, while a 2D game like *Hades* interrupts its gameplay with elaborate text exchanges that characterize its flirtatious cast. Artists interested in writing for the broadest possible audience while having the freedom to dabble in multiple genres—espionage, cyberpunk, swashbuckling, and even romance—should strongly consider the action/adventure space.

Open World

The open-world game is a recent genre that wasn't technically feasible until consoles and PCs could reliably generate enormous 3D spaces. Instead of dropping players into the cramped 3D hallways of the Shadow Moses base in *Metal Gear Solid*, players now have all of Manhattan to explore in *Spider-Man: Miles Morales* or huge swaths of Siberia in *Rise of the Tomb Raider*. It's important to recognize that "open world" is an amorphous canvas that could support action games, RPGs, and many different types of games. This murky line separating genres should be familiar to creative writers. Are the distinctions between a first-person shooter open world and an RPG open world that different from the muddy barrier dividing a lyrical essay from a poem?

The open-world genre was popularized in 2001 by Rockstar Games' now-infamous *Grand Theft Auto III*. That game's missions aren't especially groundbreaking. The player assumes control of a silent criminal looking to climb the ranks of Liberty City's underworld. This means gunning down foes, robbing innocents, and fleeing crime scenes in stolen cars—nothing players hadn't seen before in earlier titles like *Driver* or *Need for Speed III: Hot Pursuit*. However, instead of the missions unfurling one after another, interrupted perhaps by loading screens or brief bits of narrative, *Grand Theft Auto III*'s missions are scattered across a persistent map of Liberty City—a tableau of Manhattan. When missions end, players reassume control and can take on activities at their leisure. Perhaps they'll drive across town to the next mission or wander aimlessly taking in the sights. Maybe they'll drive a cab for money or assist those in need with an ambulance. Maybe they'll steal a tank and blow up some police cars. Open-world games and their illusion of freedom were revolutionary in 2001 and remain a dominant AAA genre even today.

For over twenty years, open-world games have generally followed the formula of *Grand Theft Auto III*, operating under the premise that bigger is better. Although *Grand Theft Auto III* gives you a small city to explore, its sequel—*Grand Theft Auto: San Andreas* released three years later—provides an entire state. Download a modern open-world title like *Assassin's Creed Odyssey* or *Horizon Forbidden West*, and you might feel overwhelmed by the sheer size of the map, not to mention the dozens upon dozens of marked objectives and story beats to explore. Although some games shy away from this Super Size approach—especially Japanese open-world games like *Elden Ring* and *The Legend of Zelda: Breath of the Wild*—many Western open-world games tell immense branching stories that can play out dozens of different ways based on the decisions of the player.

Take, for example, 2011's landmark open-world game *Skyrim*. It begins the same way every time—the player finds themselves arrested after entering the fantasy land of Tamriel. Moments before they're executed for their crimes, a dragon attacks, and the player escapes. What happens next is up to you. The player selects their race, their gender, their appearance, and is then confronted with a seemingly endless parade of choices. Should they follow some rebels to the next small town or the Imperial soldiers? Are they better off ditching them altogether and sticking to back roads? Should they pursue the main thrust of the story—confronting magical dragons—or should they avoid it altogether, enrolling in a magician's college or a thieves' guild? Maybe they just want to build a cottage in the woods, get married, and raise some children, spending their days chopping wood while totally oblivious to the struggles of Tamriel? Sal often teaches *Skyrim* in his classes and asks students after playing eight hours to chart their experiences. Each time the class discovers that no one path traveled was the same, yet each player was presented with their own narratives with individual concerns, climaxes, and characters. Freedom and scope are the hallmarks of the open-world genre, and this is why these games tend to be written by large teams of writers. The writing team must not only try to anticipate the disparate choices millions of players might make, but they must also shape these choices into something resembling a cohesive story.

Narrative tropes are nearly impossible to pin down for open-world games, because open-world games typically take existing genres—action, RPG, first-person shooter—and spread them out over a large map the player can explore at their leisure. While titles like *Red Dead Redemption* and *Ghost of Tsushima* take their cues from history—not to mention a liberal dose of Sergio Leone and Akira Kurosawa —games like *The Witcher 3* and *Dishonored 2* explore the kind of high fantasy we typically see in RPGs. For every *Cyberpunk 2077* that explores a high-concept sci-fi future like in a William Gibson novel, there's a *Mafia II* that unpacks the gritty underworlds of Scorsese and Coppola. Writers who want to join a team crafting dozens of detailed stories that might not even be discovered by your average player should strongly consider the open-world genre.

Walking Simulators

While open-world games are set on massive maps that take hours to traverse, walking simulators usually feature deliberately limited environments—say a house or an office building—that would appear small even compared to the narrow hallways of *Metal Gear Solid*'s Shadow Moses base. Generally, walking simulators forego what most people might consider the fundamentals of gameplay. There's almost never combat, and the player can usually only interact with the world by moving around the map while inspecting items. They might stumble across written diary entries or audio logs, slowly piecing together events through what is often referred to as environmental storytelling. In a game like *What Remains of Edith Finch*, the player takes control of the titular Edith as she returns to her childhood homestead and reflects on the many deceased members of her family. The game interrogates notions of generational trauma through a series of small curios that showcase each relative's final moments in life. Conversely, a game like *Everybody's Gone to the Rapture* opts for a more genre-heavy approach as the player explores a small town whose inhabitants have collectively disappeared. Recordings and hints are scattered across the landscape, and this is the general setup of most walking simulators—a cheeky name that addresses the genre's so-called lack of traditional gameplay. While narrative often takes a backseat to gameplay in genres like action/adventure or even the RPG, story is the fundamental attraction of walking simulators.

The walking simulator emerged in a far different fashion from the other genres we've discussed so far. Instead of beginning with development studios or game publishers looking to turn a profit, the walking simulator started with fans. Most consider 2008's *Dear Esther* as the origin of the genre. Unlike commercial releases we've discussed so far, *Dear Esther* was released as a free modification to *Half-Life*. These fan games—typically called mods by the community—became extremely popular in the 1990s. Determined fans would hack into their favorite PC games and make their own changes—say, adding Barney the Dinosaur to 1993's *Doom*—before releasing these mods online for free. Soon enough, modding became so popular that PC games shipped with tools to create your own levels and mods. *Dear Esther* emerges from this milieu, only instead of relying on the combat that is the bread and butter of *Half-Life*, it instead focuses on exploring an abandoned island while listening to a man's letters to his deceased wife. There's no combat to speak of, and, instead, the game's tone is bittersweet. This mod won the 2009 award for the Best World/Story at the IndieCade Independent Game Awards and eventually saw its own commercial release in 2012.

Dear Esther was the first walking simulator, but it wouldn't be the last. One of the first of these games to achieve massive acclaim was 2013's *Gone Home* developed by the Fullbright Company. *Gone Home* begins in 1995 when Katie—a college student—returns home from studying abroad. She goes to visit her sister and parents—who moved into their uncle's rickety mansion while she was away—but finds each of them missing. Initially, *Gone Home* misdirects you into thinking it's a horror game. Lighting crackles outside. Lights burn out. An ominous pentagram is left behind on the messy floor. But, in the end, there's no otherworldly horror haunting the Greenbriar home. Instead, what Katie discovers through a series of audio logs is that her younger sister Sam has come out as gay. Her parents react disastrously, and, in the end, Sam runs away with her girlfriend. The only horror here is what LGBTQ+ teens are forced to endure, and the only gameplay consists of wandering around the house listening to diary entries and looking at mundane objects of fascination.

In a medium where LGBTQ+ stories are historically few and far between, *Gone Home* became a breakout hit and, in many ways, is still the title people think of when they picture the walking simulator. We'd be remiss here not to mention the allegations facing Steve Gaynor, one of the founders of the Fullbright Company, arguing that he was abusive and misogynist to his fellow employees. How you choose to grapple with that complicated legacy is up to you, but we too have felt uncomfortable assigning *Gone Home* now that these allegations have been made public.

Three years after *Gone Home*, Campo Santo published *Firewatch*, easily one of the finest walking simulators yet released. While most entries in the genre take place in extremely confined environments, *Firewatch* opens that up, giving the player control over a fire lookout named Henry as he explores dense woods. Although the game tips into genre-laden cliché in its final fourth, much of the game unfolds while Henry chats on his walkie-talkie with the unseen Delilah. The player can choose their responses occasionally, but no matter what options you select, the narrative eventually reveals that Henry has exiled himself to the woods after his young wife developed early-onset dementia and is put into a care facility. The narrative that unfolds is painful yet hopeful, and the relationship between Henry and Delilah grows into something shockingly complicated. *Firewatch* is a quiet game not unlike the fiction of Rick Bass or Bobbie Ann Mason.

Although many elements of the walking simulator—especially audio logs—have been incorporated into other genres, it remains perhaps the best option for writers who want story to dictate gameplay and not vice versa. It's a perfect genre for writers looking to dabble in video games who perhaps lack familiarity with the medium, and it's also the genre that most easily syncs up with literary fiction. Walking simulators are also wonderful for players who don't have years of experience playing deceptively difficult games. When we assign a mechanically complicated game like *Skyrim*, students who struggle with games often experience little beyond the earliest moments of the game. Assign a game like *What Remains of Edith Finch*— where there are no fail states and anyone who plays long enough will complete it—and those same students light up, delighted by their own progress and the surprising depth of the narrative.

Story Platformers

The 2D platformer dates back at least to 1980's *Donkey Kong*. Here, Mario must traverse four single screens while running and jumping to avoid obstacles. 1982's *Pitfall!* in many ways popularized the genre and perfected it. In that game, the player steers an Indiana Jones-esque character who moves from left to right, leaping over pits and alligators as the screen scrolls behind them. This deceptively simple formula can be seen in games like *Super Mario Bros.*, *Sonic the Hedgehog*, and more modern games like *Shovel Knight* and *Gato Roboto*. However, all of these games are fairly light on story. So, what even is a story platformer?

A story platformer is a loose constellation of games that typically marry 2D side scrolling mechanics to extended narrative beats. The amount of story can vary. Is *The Legend of Zelda II: The Adventure of Link* with its many towns to explore and people to chat with a story platformer? Is *Prince of Persia* with its rotoscoping techniques that made its 2D characters resemble actors? We would argue yes. This term first grew popular with the 2008 release of *Braid* and has been applied retroactively ever since. In that game, written by Jonathan Blow, you steer an avatar through a series of Mario-style levels. Except here you're presented with text scenes in between levels that reveal a surprisingly bittersweet story about male entitlement that eventually intersects with the gameplay itself—represented by the player's ability to reverse time and undo their mistakes.

A year later, Paolo Pedercini, a professor at Carnegie Mellon, released *Every Day the Same Dream* for free online. This game sees the player assuming the role of a white-collar worker drowning in the repetition of their humdrum life. If the player follows the obvious path and heads from their apartment to their office, the game restarts. It's not until the player begins deviating from that path that they begin to experience the game's deeper anticapitalist message. Another lovely example is *Celeste*, written by Maddy Thorson in 2018. This game begins unassumingly enough when the player is tasked with guiding the titular character up a mountain. However, as the game proceeds, the player must contend with a literal manifestation of the protagonist's insecurities and depression. At one moment, you must even help Celeste focus on her breathing during a panic attack.

What do all of these examples demonstrate? Most story platformers rely on very limited controls—generally moving from left to right and jumping—and genre tropes lifted from *Super Mario Bros.* that nearly anyone who's ever seen a video game will immediately comprehend. They're simple enough that they can often be designed by one person or a very small team and that allows these games to venture narratively into spaces that bigger budget games with massive corporate structures and stockholders wouldn't dare to go. Remember the narrative concerns of the three previous examples—male entitlement, anticapitalism, and depression/anxiety—and compare them to more big-budget games like *Uncharted 2* or *Ghost of Tsushima* or *Shadow of the Tomb Raider*. The latter games may occasionally dabble in serious themes, but, at the end of the day, their goal is to entertain the largest number of players possible. Story platformers—and especially free ones like *Every Day the Same Dream*—aren't like this, and the analogy we use with our students is to think of story platformers and other indie games like indie films. These are the video game equivalents of *Minari* or *She's Gotta Have It* or *Slacker*. The big-budget games are more like *Furious 7* or *Avengers: Endgame* or *Jurassic Park*. It's up to you to decide which sphere you want to write in, and it's critical that you consider genre when you're deciding how personal a story you wish to tell.

Visual Novels

When we're asked about the easiest path for creative writers familiar with traditional genres—fiction, nonfiction, and poetry—to move into writing video games, visual novels are the obvious first answer. The term "visual novels" alone tells you nearly everything you need to know. These are long-form stories paired with images that a player clicks through a screen of text at a time. Sometimes the player makes a choice and the narrative branches like in the choose-your-own-adventure books you may remember from childhood. Sometimes the player may have to simulate a dice roll like in tabletop role-playing games. But generally, visual novels are the genre most concerned with text and creative writing. That makes them an easy starting point for writers.

In 1983, developer Chunsoft released *The Portopia Serial Murder Case* for the NEC PC-6001 computer. Unlike most games from that period, the player isn't given one-to-one control of an avatar. Instead, to solve the game's mystery, you must read dense pages of text while making cerebral choices. This is what made *Portopia* stand out, attaining mass popularity in Japan where it was ported to multiple platforms, eventually selling over 700,000 copies on the Famicom—Japan's version of the Nintendo Entertainment System. A host of visual novels appeared in the wake of *Portopia*'s success, including the much-lauded *Snatcher*—a *Blade Runner* homage—from *Metal Gear*'s Hideo Kojima just five years later.

Historically, the visual novel was far more popular in Japan than anywhere else in the world. In the 1980s and 1990s, these games were rarely translated outside of Japan, and a number of diverse genres inside the visual novel flourished. Most prevalent were dating simulators, usually set in high schools, where players explore one romance after another. The most famous example is the Tokimeki Memorial series which is so popular in Japan that it spawned forty-six different entries, a manga, an anime, and a live-action film—none of which have ever been commercially released in the United States. Aside from romance, sci-fi and fantasy are the biggest visual novel genres. Slowly, these games are spreading outside of Japan. In 2022, SquareEnix finally released 1996's *Radical Dreamers*—a direct continuation of their 1995 JRPG *Chrono Trigger*—in North America. And with the advent of digital distribution, many Japanese game publishers have begun releasing their latest visual novels electronically. Notable examples include *The House in Fata Morgana* and *Danganronpa*.

Western developers have recently flocked to the visual novel, thanks to the Ren'Py Visual Novel Engine. We detail how to use it in Chapter 12, but this easy-to-learn development tool allows writers to assemble visual novels for commercial release. Two examples of Western visual novels worth noting are *Citizen Sleeper* and *VA-11 Hall-A: Cyberpunk Bartender Action*. The former is an anticapitalist screed that uses the cyborg trappings of *Snatcher* and *Blade Runner* to tell a story about corporate greed and exploitation, while the latter sees the player manning a dystopian dive bar and chatting up regulars. This cross-cultural conversation in many ways mimics the proliferation of the JRPG.

For any writer interested in games who isn't ready to dive headfirst into more mechanically complicated genres or design suites, the visual novel is a lovely place to begin. All you need to learn is Ren'Py, and if you know how to draw or compose music, even better. If not, find some collaborators. The visual novel's limits are as vast as your imagination and can just as easily support romance, horror, realism, sci-fi, or any other story you wish to tell.

Exercise

(1) Choose any fiction you know well that isn't a video game. This can be a novel or a short story or a movie or a TV series, for example. Then, select the video game genre from the list above that you know best. Write a two-paragraph summary of how you would translate

one of your favorite pieces of fiction to one of your favorite video game genres. What might Toni Morrison's *The Bluest Eye* look like as a walking simulator? How about *Euphoria* as a role-playing game or *Lady Bird* as a visual novel?

(2) When you're done, take that same fiction and now write another two-paragraph summary explaining how you'd translate that work into the video game genre above that you're least familiar with. What must change in the shift from open world to story platformer, for example? After both summaries are completed, reflect on the ways in which the rules of genre dictated how you told the exact same story.

8

Game Design and the Creative Writer

A Rationale for Game Design

Dontnod's 2015 game *Life Is Strange* opens with a photography student, Max Caulfield, taking a break in her high school's bathroom when two central characters, Nathan and Chloe, burst into the room in the midst of a heated argument. They don't notice Max, and their fight escalates, until Nathan shoots Chloe in the stomach in a fit of anger. When Max steps forward, raising her hand in terror and shock, she suddenly finds herself back in her photography classroom, a few minutes before the murder, astonished at her newly discovered ability to rewind time and desperate to save Chloe from murder. On the surface, the power of this opening might seem to stem from its story—relatable characters, conflict and homicide, a magical superpower, and a high-stakes goal with a ticking clock. It's true that the story in *Life Is Strange* successfully relies on many traditional creative writing approaches, but focusing on the strength of its story only explains part of what makes this opening so riveting.

If you were asked to write the script for a movie, basic creative writing skills would help you craft many important elements—strong characters, an engaging plot arc, a compelling theme—but without a knowledge of the tools specific to film—things like music, lighting, camera angles, and editing—you wouldn't be able to take advantage of the full power of the medium. The same is true for games. While narrative skills are certainly helpful in creating video game stories, in order to make your games as engaging as possible, it's important to also have an understanding of basic game design principles. What is a game, and what are some things to consider when creating one? How can game design and narrative work together? What kinds of effects are possible in games that are difficult to replicate in other media? What can you do to make the most engaging game experience possible? The next few chapters will focus on elements of game construction and game design and explain how a knowledge of these concepts is useful for creating engaging game narratives.

The following pages will give a rundown of important game design concepts that are particularly useful in the context of writing strong narrative video games. Game design is a complex and fascinating field, and we won't have enough space to cover everything in this brief chapter. If you're interested in writing games, it's a great idea to spend some time familiarizing yourself with these concepts and learning more about them. Many of the ideas in this section are adapted from the following books: *Rules of Play: Game Design Fundamentals* by Katie Salen Tekinbas and Eric Zimmerman, *Game Design Workshop: A Playcentric Approach to Creating Innovative Games* by Tracy Fullerton, and *Games, Design and Play: A Detailed Approach to Iterative Game Design* by Colleen Macklin. As we will explain in this chapter, understanding basic game design will not only help you create engaging games, but it can strengthen your creative writing in other genres too!

Boundaries, Constraint, and the Magic Circle

Imagine you are participating in an in-person social deduction game like *Werewolves* (or perhaps *Mafia* or *Secret Hitler*), where some players are secretly assigned the role of assassins and must win the game by killing other players. The rest of the players must try to determine who the villains are and vote them out before being killed themselves. When you're playing this game, are you worried that your fellow players might become confused and actually kill you? Is it possible to get confused about the boundaries of the game when you are the killer yourself? If your friend betrays you in the game, does the sting of that duplicity carry over into the rest of your life? The answer to most of these questions, of course, is no. While in-game social encounters may sometimes result in hurt feelings or long-term resentment, for the most part, we recognize that game spaces are separate from everyday life, and we don't take it personally when our friends betray or lie to us (or even kill us). And while we might reminisce about things that happened in games or tell funny stories about past game experiences, and though these memories can be joyful and meaningful, we understand that game spaces are different from the spaces of everyday life. When a friend lies to us or steals something from us in a game, this feels very different from when someone lies or steals outside the game, in day-to-day life.

In his classic text *Homo Ludens*, Johan Huizinga describes this effect as a magic circle, or a social space that operates under different rules and expectations from nongame spaces. While the specifics of these spaces vary from game to game, for the most part, this magic circle is reliant on boundaries, rules, and shared social agreements. Take a game of soccer, for instance. Even though it would be much easier to throw the ball or carry it, soccer players (aside from the goalie) agree to a key rule—that they will not use their hands or arms to move the ball. They understand that the game is played within the confines of a specific space (so they don't go running off into the bleachers, for instance), and they agree to basic rules of social conduct and refrain (mostly) from things like punching or tackling one another. Huizinga argues that the magic circle means that games are a space apart from everyday life. They require players to agree to abide by specific constraints, to uphold boundaries (both physical and social), and to work together to maintain, navigate, and discuss these elements when there are conflicts or disagreements. As designers, it's up to us to consider the boundaries and rules of the games we craft and to articulate our visions clearly, but players carry a great deal of responsibility, too, in creating and maintaining the experience of a game.

From a writer's perspective, the concept of the magic circle is useful in a few ways. First, it reminds us that game design elements like rules, boundaries, and constraints can never be completely separate from the stories we create. In *Werewolves*, for instance, one person often acts as a narrator, crafting a spooky story about werewolves terrorizing a village while also managing important game elements like voting and negotiation. The more effective the narrator's story is in integrating game elements seamlessly, the stronger it is and the more it helps to reinforce the social contract of the magic circle. When crafting narratives for games, it's useful to remember that stories that work with and support game rules and mechanics are often stronger than those that do not, and in many ways, crafting game narratives and refining game rules and mechanics go hand in hand. As a game writer, you might ask yourself: What rules must players follow in this game, and how can my story help to support and work with those constraints? What kinds of player actions are possible, and how can my story help to explain and emphasize those possibilities? What are the boundaries of this game, and how can the narrative help to explain those limits?

Even more importantly, an understanding of the concept of the magic circle gives us tremendous power as writers and helps to explain why game narratives can be particularly compelling. Huizinga and other scholars have noted that the experience of the magic circle as a "space apart" means that players are open to sitting with ideas, concepts, and actions that might be difficult in other environments. Because of the boundaries of the magic circle, you might feel a freedom to say things in a game of *Werewolves* that you wouldn't otherwise. For instance, you might comment on a friend's style of communication. ("You always tell people what they want to hear.") Or you might draw awareness to social hierarchies that would otherwise go unchallenged. ("Usually we let Grandma decide, but what if she's a werewolf?") The magic circle allows such observations to occur without the social repercussions that might happen in other circumstances. As writers, it's helpful to think about how the game narratives we create might be useful for encouraging players to engage with or consider existing social structures, and how a story might help people to think in new ways about familiar experiences and relationships.

Further, because games are interactive, the weight and effect of certain concepts and ideas can be even stronger, and players tend to be more willing to engage with challenging ideas while in the relative safety of the magic circle. In Tracy Fullerton's *Game Design Workshop*, game designer Brenda Romero tells a story about speaking with her young daughter about the Middle Passage and being surprised at how unaffected her child seemed to be, despite her familial connections to it. It wasn't until Romero and her daughter created and played a simple game about the Middle Passage, where they were forced to make decisions based on things like the capacity of the boat and the amount of food available, that her daughter seemed to grasp the emotional impact of the Middle Passage and to begin to understand the weight of the experience. Because of the relative safety of the magic circle and its experience as a space apart, it can act as a way of defamiliarization by encouraging players to engage with well-known ideas and topics in new and unexpected ways. Further, because most games require players to act and to make choices, participants may engage with these ideas on a deeper, more personal level than more passive media experiences like films or books.

As writers, it's important to keep the power of the magic circle in mind, as it is one of the most compelling reasons to tell a story in a game format. What kinds of subjects might be particularly useful to engage with in an environment that exists as a "space apart" from society? How might collaborative game storytelling experiences be particularly useful in exploring a certain topic? What kinds of preconceived ideas or implicit biases might players bring with regard to a particular topic, and how could a game experience encourage them to engage with a subject differently or to approach it from a new mindset?

To return to the example of *Life Is Strange*, part of the success of its opening lies in its use of the magic circle. Players learn right away that they must solve the problem within the physical boundaries of a few rooms of a high school, that there is a time limit, but that they are allowed to engage in conversations and then replay them by rewinding time until they have achieved a satisfying result. To continue with the game, players must save Chloe, but succeeding in doing so is completely reliant on the player's choices, providing them with a great sense of responsibility and agency. Further, to save Chloe, players must begin to understand a crucial component of gameplay—that social interactions are important, and that paying close attention to character details and the environment leads to stronger results. Thus, within a few minutes, *Life Is Strange* establishes a magic circle where players understand the importance of building character relationships and managing interpersonal conversations. This opening encourages players to empathize with characters and teaches them the rules and expectations of the game, all of which helps to deepen player connection to the story.

Meaning and Experience

One of the most complex things about writing for games is how difficult it can be to predict player behavior. In *Life Is Strange*, for example, after Max learns about her new superpower and Chloe's imminent death, some players might hurry to the bathroom as quickly as possible, eager to save Chloe, while others might choose to wander around the photography classroom, exploring all the small details of the world and searching for secrets. Even in a somewhat structured setting, like Max's photography classroom, it's impossible to predict where players will go first. Will they chat with a classmate or talk to the teacher? Will they examine the posters on the walls or ignore them altogether? Unlike a film or a novel, where the story proceeds in the same way each time, games rely so much on interactivity that it's rare for a game experience to proceed in the same way twice. Indeed, part of what makes a game narrative so satisfying is the sense of being able to engage with the story and make it your own.

As a writer, though, telling a story in such an unpredictable medium can seem challenging. How can we make sure that players are finding the most important elements of a story and that they're encountering them in a useful order? How can we allow our story to be fluid and encourage player choice and interaction while still making sure that it makes sense? We'll talk more about things like designing choices and encouraging replay in the next chapter, but before we get to specific structural considerations, let's start with something simple. As much as we may want to create player experiences that are completely predictable or to ensure that everyone who engages with our game does so in a foreseeable way, this is impossible. In fact, some players might even go out of their way to try to do something strange or humorous. In the game *Firewatch*, for example, most players will explore the forest and follow the story line. But some players will try to finish the game as quickly as possible (in barely an hour), while others might decide to collect all the books or to see what happens if they try to travel to an unexplored part of the map. As game writers, we need to make our peace with this. Not only is it infeasible to create an entirely scripted experience, but it also goes against the spirit of many game narratives, where part of what is most enjoyable is a sense of player responsibility and agency.

This doesn't mean, though, that as game writers we must relinquish authorial control entirely. Instead, it's useful to approach game writing from a different perspective, one that embraces open-endedness and unpredictability. One of the most useful places to start when writing a game is to think about meaning and experience. While you may not be able to predict exactly how your story will unfold for each player, you can make design choices that help to give a sense of continuity and connection between all of the encounters a player has in the game, regardless of what they are and in what order they happen. First, as a game writer, it's useful to start by thinking about the kinds of experiences you would like for players to have within the game you are creating. Are you crafting a world that is dark and gritty, one with severe life or death consequences? Or are you constructing something humorous and light-hearted, a place where opposing choices may have equally positive outcomes? How do you want your players to feel? What kinds of things do you want them to pay attention to? What kinds of consequences should be possible within this world? Thinking carefully about the game experience you'd like to create is similar to establishing the tone or mood in a story, and once you've decided on it, it will influence all the artistic choices you make in your game—including language, worldbuilding, visual style, and more. What's more, if you're creating a game as a team, deciding together on the kind of experience you'd like to create from the beginning can make it easier to coordinate choices, so that all the elements fit together even if more than one person is creating them.

Let's return to the example of *Life Is Strange*. In this game, players are meant to carefully consider their interactions with other characters, so although it might seem like an unrealistic departure from the story for Max to be able to spend time talking to her classmates or examining posters when Chloe is in danger, allowing players to take these detours encourages them to focus closely on setting and character, which helps to establish a crucial experience of the game world right from the beginning. Similarly, the central game mechanic—being able to briefly rewind time—helps to create a world where players are rewarded for taking their time and considering their choices carefully. Though there are a few moments in the game with stressful time constraints, for the most part, this game encourages players to slow down, to enjoy nature, listen to music, reflect on the past, and consider art, all of which are useful strategies in generating empathy, which is also an important experience of the game.

Beyond creating continuity through game experiences rather than a rigid plot order, it's also useful to consider the concept of meaningful play. In *Rules of Play*, Salen and Zimmerman explain that the concept of meaningful play is a core element of their approach to game design, arguing that "the goal of successful game design is the creation of meaningful play." Meaningful play is a complex concept, but at its most basic level, meaningful play relates to the actions that are possible within the game world as well as the larger "meaning" of the world overall. In the next chapter, we'll talk in more detail about ways to approach writing meaningful choice, but as you're designing the kinds of experiences you'd like for players to have in your game, it's important to think about what kinds of actions and choices you'd like for players to make and how you will communicate the meaning of those choices to players. In *Life Is Strange*, for instance, helping Max to create close connections with those around her is a key part of the game experience, and rewinding time allows players to see how different responses change the result of a given conversation. For instance, after saving Chloe, Max has the choice to report Nathan to the principal or hide the truth. Players have the option to play out both versions of the conversation before deciding on their final choice, and the story continues either way, albeit with different consequences. The game teaches players to find meaning in discussions that are thoughtful, empathetic, and kind, and while it may not always be possible to predict the outcome of a particular choice, the feedback that players receive after making choices that honor those values makes it clear that this approach holds meaning within the game.

In addition to considering how meaning works within the world of a game, it's also useful to think of meaning more broadly in terms of what a given narrative might mean for a player. What would you like for players to take away from your game? How would you like for them to think differently about themselves or the world? How might this story change how players think or act? In what ways might the experience of this game be transformative for players? One of the most powerful things about game narratives is the way that they can encourage players to think in certain patterns or to approach things in specific ways. As game writers, we're asking players to act, to choose, to be conscious agents rather than passive observers, and one of the consequences of this kind of storytelling is that it can train players to look at the world differently, even after they've left the environment of a particular game. If *Life Is Strange* rewards players for approaching conversations with other characters in ways that are open, accepting, and curious, it helps to encourage people to find meaning in these kinds of interactions outside of the game world as well. In other words, it helps to create meaning by drawing awareness to elements of the world that exist beyond the game and encouraging players to think in ways that may continue even after the game is over. While this kind of writing is certainly possible in other media, it is a crucial element of writing a game-based narrative and one reason why someone might choose to write a game to tell a particular story.

Case Study: *NieR: Automata* (2017)

No one would blame you after playing the first half-hour of *NieR: Automata* for assuming it has no interest in storytelling whatsoever. The game opens with the briefest of voice-overs before dropping you into an intense shooter where the player controls a jet on a 2D plane, harkening back to games like *Asteroids* or even the aforementioned *Spacewar!*. Before long, the game shifts into a 3D action landscape where you control 2B, an android, who's sent to Earth to aid a resistance fighting armies of mindless machines. Even this isn't especially promising at the narrative level and reads like a ho-hum and unassuming story meant to motivate the player through dozens of hours of combat. However, *NieR: Automata* is far more than just another soulless action game meant to pass the time.

NieR: Automata is directed by Yoko Taro, who is considered by many as the new maestro of meta games. Taro remains a mysterious presence in the gaming world, similar to a writer like Elena Ferrante who doesn't appear in public and uses a pseudonym. Unlike Ferrante, Taro has been photographed, but only while wearing an elaborate mask of a sneering moon that covers his entire face. This affectation is no mere publicity stunt meant to cover up any narrative shortcomings in Taro's work. Very quickly, *NieR: Automata* complicates its plot, exploring themes of pacifism, free will, spirituality, and the relationship between man and machine. By 2017, many of these themes may have felt pat or redundant, done better and earlier by games like *Metal Gear Solid* or by anime like *Ghost in the Shell*. After twenty to twenty-five hours, however, the game reaches its conclusion and that's when the deeper experience of *NieR: Automata* is revealed (Figure 10).

After the credits roll, the game immediately restarts, only this time the player isn't controlling 2B. Now, the player finds herself guiding one of those "mindless" machines they've been murdering for hours upon hours before gaining control of one of 2B's companions. The title flashes across the screen, and you replay the entire game again from a different perspective, learning crucial new elements about the world and its characters, similar to the nonlinear storytelling of a film like Akira Kurosawa's *Rashomon*. Another five to ten hours

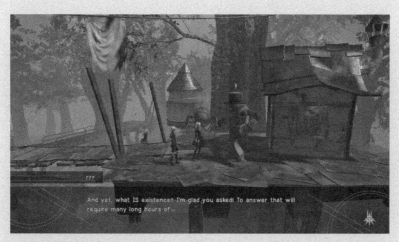

And yet, what IS existence? I'm glad you asked! To answer that will
require many long hours of —

Figure 10 Machines contemplate the nature of existence in 2017's
NieR: Automata.

pass, and the player completes the game a second time, and guess
what happens? It restarts again.

All told, *NieR: Automata* has twenty-seven different endings
stemming from five different routes that tunnel through the game.
While the overall story may replicate beats we've seen before, it's the
structure in *NieR: Automata* that is revelatory. Think of *NieR: Automata*
as a postmodern wiki or hypertext. It begins with a narrow path
pushing the player toward certain outcomes, but very quickly that path
branches again and again shuttling you toward radically different
endings, an electronic *Infinite Jest* that becomes denser and more
complicated with each new ending and unreliable retelling.

The ambition of *NieR: Automata* becomes clear in the game's
"final" ending. Here, the player assumes control of a ship on a 2D
plane—much like in the opening—only now they're being attacked by
the literal end credits of the game. The many names of the dev team
fly past, shooting waves of colorful bullets, and at first most players
will be able to survive. However, the screen soon becomes so chaotic
and cluttered with neon death that the player has no option but to
accept an offer of help. Suddenly, the player is flanked by other ships,
not to mention words of encouragement written by other real players
from across the globe. The game's overriding narrative concerns—
collaboration and hope amid the collapse of society—merge with
gameplay as the player must rely on others to survive this final barrage.
The player is then presented with a truly meaningful choice. Video

game choices before this were purely hypothetical—making small or large changes to a digital world outside of our own experience—but *NieR: Automata* gives you the option to write an inspirational message to future players while transforming into an AI ship that will assist them during this final sequence. However, to help these random players you've never even met, you must erase your own save file.

Let us be clear. To reach this point in *Nier: Automata* will take an average player forty to sixty hours, and it would be almost impossible to see all twenty-seven endings in that time. While most video games encourage a maximalism where you must see and do everything, *NieR: Automata* zags, asking you to sacrifice your own progress to assist others. It's a deeply personal choice only possible in video games, and it's complicated by *NieR: Automata*'s obsession with structure, repetition, and exponentially branching paths. *NieR* doesn't merely replicate the tropes and plot devices of old media like literature or film. Instead, it considers the new possibilities video games afford writers and hardwires that potential into the thematic core of the game. *NieR: Automata* is a game that is as concerned with what it means to tell a story in a video game as it is concerned with shooting machines. That's progress.

Choice and Action

As we've mentioned earlier in this chapter, interactivity is a crucial component of game design, and allowing players the freedom to make choices and take action within a narrative is one of the most important elements of successful game storytelling. A useful place to start as a new game designer is to consider the actions and choices that are available to players in your favorite game. What can players do within a game? How are their actions limited? What kinds of choices are possible? In *Life Is Strange*, for instance, players can walk freely within each area, examining objects and talking with the people they encounter. They can interact with the world, but only if there is a specific indicator in the environment that indicates this is possible. They can take photographs and access information such as Max's text messages and journal. In most conversations, players are able to choose from a variety of responses and are even able to rewind a conversation and replay it differently once they've learned a crucial bit of information, but once a conversation is complete, they must live with the narrative consequences of their choices. There are also a number of things players cannot do in the game. For instance, they can't write in Max's journal or respond to text messages. They can't leave the confines of a particular environment until they've completed the story objectives, and though it is possible to respond harshly to someone verbally, in general, the game doesn't allow physical interactions, like punching or kicking someone.

As a game experience, *Life Is Strange* clearly prioritizes empathetic conversations with other characters and close attention to the world, so the decisions about the kinds of choices and actions in the game make sense. Allowing physical interactions would distract from encouraging close attention to discussions with other characters and because the game helps to draw attention to the way small choices in dialogue responses can lead to large impacts, emphasizing choices in conversation experiences helps to support that. Including interactable elements within the game world helps to encourage players to examine the environment closely, and the photography mechanic trains players to appreciate the artistic nature of their surroundings while also serving as a device to help players remember what has happened in the game so far.

Unlike in books or film, where it is easy to reread or rewind if participants want to review something that has happened, if players want to revisit a past game experience, this often requires replaying, which can be tedious. So often games will allow actions that help to remind players of narrative elements, like Max's journal, text messages, and photography album. It might seem strange not to allow players to choose responses to text messages in a game that is meant to encourage players to think carefully about specific elements of interpersonal interaction, but limiting this action helps to keep the choices from being overwhelming while also allowing the game writers some control over the story. The text messages help to remind players of who the characters are and what Max's relationship is to them, and the journal creates a running textual record of the story that players can return to at any time if they forget what is happening or what they are supposed to do. While these elements are useful, they are also optional, and if players choose to ignore them completely, this doesn't really affect the game experience, though the photo album element might encourage players to explore the world thoroughly to collect all the pieces.

When designing a game, it's important to think about the role of choice and actions within the larger context of the meaning and experience you'd like to create for players. What kinds of choices would you like for players to make in your game? How will you communicate the effects of those decisions? What kinds of actions will you allow players to take, and how will you help them to understand what consequences might come from those actions? Similarly, what kinds of actions and choices would you prefer to leave out of your game? What kinds of behaviors and approaches don't fit with the vision you have for this world?

Lack of choice, too, can create a strong narrative impact. For instance, in Zoë Quinn's 2013 game *Depression Quest*, a text-based game meant to help players understand the experience of depression, choices may be grayed-out or unavailable to players to highlight how depression can make actions like leaving the house or calling a friend difficult. Similarly, Nina Freeman's 2015 game *Freshman Year*, a diary-style game about sexual assault, allows players to make choices about what to wear and how to behave while out at a college nightclub, but regardless of what choices the player makes, they experience an assault near the end of the game. By making it impossible for players to avoid this, Freeman highlights the character's lack of agency and power while also critiquing the social tendency for victim blaming.

While it's important to note that to some extent, choices and actions are determined by the game writing program you're using, it's crucial not to let preconceived ideas about your abilities or the capacity of the software you're using keep you from making conscious decisions about choice and action within your game. We talk more extensively about tool suites later in the book, but if you're making your first game in Twine, your narrative may be largely text-based, and you may use default settings for things like font, background color, and text alignment. Similarly, if you're working in Bitsy for the first time, you'll probably rely on simple visuals and short pieces of text. When creating any game, it's useful to ask yourself questions about what is possible (both within the program you've chosen and with your own skills and abilities). How can you make the most of the Twine text-based format to highlight the kinds of actions and choices most important to your game? How might the nostalgic feel of Bitsy work with the overall mood you'd like to establish in your story? What kinds of tools are available to you as a writer working to tell a story in a particular form, and how will you make the most of them?

Resources, Scores, and Rules

As we've discussed, game experiences rely on rules and constraints, and often these rules are connected to resources or tools. These kinds of tools can take many forms, but often they limit the actions a player can take, or the choices possible, or operate as a scoring mechanism that helps players evaluate the results of their choices. In Inkle's 2014 text-based steampunk adventure game *80 Days*, for instance, players are asked to take on the role of Passpartout and to care for their employer Phileas Fogg as he attempts to circumnavigate the globe within eighty days to win a bet. Throughout the game, players make choices about where to go and how to travel, as well as whom to talk to and what kinds of responses to give in various conversations. The choices and actions a player can make are closely linked, though, to the resources available in the game. Passepartout must pay for things like lodging and transportation, and so money is one of the most important resources players must manage on their travels. Fogg's health can decline if travel and lodging conditions stress him, while rest and care from Passepartout can help recover it. Fogg's relationship with Passepartout can deteriorate or improve based on their interactions, and Passepartout can develop certain personality traits depending on his choices. Collecting and selling items helps players to make money on their travels, but due to luggage and transportation constraints, players must think carefully about which items to take with them. Further, the constantly ticking clock of the bet creates tension as players rush to finish the journey within the expected time frame and win.

80 Days is largely a story-based game, but these resources add a larger element of choice to the narrative, as players are forced to consider the impact that their choices might have on game elements such as money, time, and health. From a game design perspective, this draws players into the game, and provides a variety of elements that players must manage and consider when making choices. From a writer's perspective, though, game resources and rules present several challenges that are important to consider when designing your story. First, what resources will play a role in your game, and how will you explain to players what those resources are and how they work? Some games may include instructions or tutorials that help to explain the important rules and elements of the world, but many players prefer to skip introductory materials and get right into the game. As a writer, it can be useful to consider working explanations of resources into the story. In *80 Days*, for instance, it's easy to understand resources like money, time, and health in the context of the story, but writers must also make it clear how choices within the game will impact those resources. This is particularly important early in the game when players are trying to get their bearings. How can your story help to explain your game's resources and tools? How can you help communicate to players how their choices might impact game elements without giving too much away or encouraging players to make choices based solely on numbers or scores? Tools and storytelling are closely intertwined, and the more your story helps to account for and explain game resources, the easier it will be for players to immerse themselves in the story.

Even more importantly, rules and scoring can help develop narrative tension. For instance, Meg Jayanth, the creator of *80 Days*, has talked about how she'd hoped in her game to pull players into the stories they encounter while traveling to help push back against some of the colonial perspectives present in Jules Verne's original novel, *Around the World in 80 Days*. In Jayanth's version, she wanted players to understand that their stories as travelers were no more important or engaging than those of the people they encountered. She hoped players would even get so wrapped up in the stories they uncovered that they might take extra time out of their travels to engage with them, even if that meant they would lose the bet. This technique for creating narrative tension is an example of a craft approach that is only possible in a game, where things like rules and resources can work in connection with the central story or create friction with it, creating an engaging story experience.

Symbols and Images

In the opening to *Life Is Strange*, when Max steps into the bathroom for a brief respite from her unkind classmates, she notices a blue butterfly in the corner of the room and stops to take a photo of it, which is why Nathan and Chloe don't notice her when they come into the room. The beauty of this moment is quickly overwhelmed by the subsequent drama, but the blue butterfly becomes a powerful symbol that persists throughout the game. First, it demonstrates the importance of noticing unusual and wondrous things, even when in the midst of difficult experiences, which is a central theme in the game. It also coincides with the moment that Max learns of her superpower and comes to represent how one small person can have a profound impact on the world around them. Further, it represents the "butterfly effect"—the idea that a butterfly flapping its wings in one place could cause a storm in another, highlighting how seemingly small choices can ripple outward, causing large-scale impacts. The butterfly also acts as a tool for telling players when a choice has been made that will impact the story, as the butterfly icon flickers in the corner of the screen when an important choice has been made that can no longer be reversed. The butterfly in *Life Is Strange* acts on multiple levels: a plot device that aids in the story, an action that trains players to appreciate beauty and look closely at details, a symbol of important themes and messages in the story, and a tool that helps players understand the constraints of the game world.

As a writer, you've probably spent time thinking about symbols in written contexts, whether it's discussing the meaning of the white dolls in Toni Morrison's *The Bluest Eye* or the strange heartbeats the protagonist hears coming from under the floorboards in Edgar Allan Poe's "The Tell-Tale Heart." When writing games, though, considering and constructing symbolic meaning becomes even more important and useful. In *Rules of Play*, Salen and Zimmerman explore the role of semiotics in game design in detail, explaining how games frequently operate using a language of signs. In a basic sense, the term "semiotics" refers to the study of signs and symbols, and it translates to one of the ways that humans are able to evaluate their surroundings quickly in a way that helps us to survive. Semiotics help to explain why we might share certain cultural assumptions about cars, for instance, so that if we see a minivan on the highway, many people might guess that it belongs to a family, and make further presumptions about things like soccer practices and social class and what that family does for fun on the weekends. In the context of game design, semiotics are useful because often signs and symbols stand in for larger concepts or ideas, and as writers, part of our task is to communicate these ideas to players.

In the context of game design, the terms "sign" and "symbol" act much more broadly than you might be used to from the study of literature. As in a story like "The Tell-Tale Heart," a symbol in a game can be an audio cue, like a repeating motif or sound that helps to evoke an emotion a character might be feeling (crushing guilt, perhaps). A symbol can also be a visual icon that helps communicate information about game resources to players, such as a heart that helps illustrate when a character loses or regains health or a coin image that indicates how much money a player has. A sign can also be an avatar or image that represents the player. For instance, Salen and Zimmerman use the example of Xs and Os in a game of Tic-Tac-Toe, which, of course, represent player identity on the claimed sections of the game grid. A sign can also be an action, such as an animation that indicates that a character is casting a spell or a visual cue that an enemy has been vanquished, or in an in-person game, a gesture might represent an important game mechanic, such as touching someone's arm in a game of tag. Symbols can act as visual thematic devices that draw player awareness to important ideas and messages in a game, and frequently, as in the example from *Life Is Strange*, symbols can act on multiple levels, working in several different ways at once.

As game writers, it is important to consider the role of signs and symbols in the games we create, because frequently it is our job to construct meaning and to help players understand the larger messages connected to the images, actions, and sounds in our games. If our game includes a digital icon of a minivan, for instance, it's important to understand that players may come to the game with certain associations, and that they may link a minivan image with the family and social elements described above without even realizing it. If we'd like players to associate the minivan with something different (kidnappers or smugglers, for instance), we need to take the time to establish those connections and develop those meanings. If we're working with a common icon (hearts representing health, for example), we may need to do very little narratively to help players understand that meaning, but if our world includes uncommon resources, we'll need to explain those symbols, particularly if we're asking players to create unusual associations with familiar items. (For instance, in one of Juli's favorite childhood games, Interplay's 1986 adventure *Tass Times in Tonetown*, guitar picks are used instead of currency.) This is particularly true when crafting avatars or central characters, where you'll often want to convey complex emotional states, important changes or realizations, or complicated backstories in a very short space.

In short, as a game writer, it's useful to think even more carefully about ways to communicate using signs and symbols, whether these are visual icons, sounds, or actions players take in your game. What resources, rules, or mechanics are present in your game, and how can you communicate details about these elements quickly? What visuals will you rely on in your game, and how will you help players understand the meaning of those icons in the game? How might you use repeated sounds or motifs in your game to help evoke complex emotions or character states? Change is an important element of most fictional narratives, and it's useful to consider what kinds of transformations might take place in your game, and how you will ask players to participate in them. How will central characters develop over the course of the narrative or how will their perspectives transform. How can you communicate these complex changes to players? As writers, we may be accustomed to relying on words to communicate, but in games we have so many more tools at our disposal. The more we develop and refine our skills for using things like image, sound, and symbolism in our stories, the stronger and more engaging our narratives will be.

Challenge, Failure, Replay, and Endings

Stories in books and films tend to proceed in the same way regardless of what we do. If we fall asleep while watching a movie, the story continues right on without us, all the way until the end. Similarly, although our understanding of the story in a novel may deepen and change the more times we read it, the story itself does not change, and we cannot alter it. In games, however, this is often not the case at all. In *Life Is Strange*, there are multiple possible endings, as well as large-scale choices that impact things like whether a character lives or dies, or the kind of relationship Max has with them, as well as multiple small-scale choices that might seem irrelevant, such as which breakfast Max chooses to order at the diner. If we fall asleep while playing *Life Is Strange*, Max will often just stand still, patiently waiting for us to wake up and continue the story. As players, this is one of the things that can make storytelling experiences in games so riveting. We're part of the narrative, responsible for the twists of the story and the ways in which a character develops and changes. Writers and programmers have created the possibilities, but it is up to us to navigate and shape them. Without us and our participation, most game stories could not exist.

As game writers, though, it's important to think about the ramifications this has for how we tell stories in this medium and to consider how design principles might impact our stories. In many games, for instance, it's common for difficulty to increase as players move through the story, so that the first levels help to establish basic skills and approaches that become more complicated in subsequent parts of the game. Take Simogo's 2019 game *Sayonara Wild Hearts*, for example. This game tells the story of a young woman whose heart has been broken and who must race through a series of encounters with tarot card arcana, battling villains while healing her emotional wounds and growing stronger. In the first level, players are asked to navigate a simple track, where they learn to collect hearts, avoid obstacles, and move in time with the music, three skills which are crucial to succeeding at the game. Each subsequent level builds on and develops these skills, so that in some levels, players learn an additional skill, like pushing a button in time with the music, while in others it becomes more difficult to avoid obstacles, or battling the villains requires more complicated sequences of motion. Part of the fun of games like this is the feeling of success and development; as we move through the levels, we become better at the game, and each level raises the difficulty just enough to challenge us while avoiding frustrating us.

Design like this requires planning, playtesting, and refining, and in many cases, strong level design is also closely linked to story development. In *Sayonara Wild Hearts*, the story of a woman overcoming a broken heart works alongside the carefully structured scaffold of the game's levels. As players, we develop our own skills throughout the game, becoming stronger and more capable, and our experience mirrors that of the central character, so that her success at the end of the story feels to some extent like our own success. Rather than watching a character overcome a broken heart, we have participated in this experience, and even though the specific details of our lives may be very different from hers, we, too, have failed and struggled, and ultimately become stronger through the game. In this example, narrative and level design are closely linked, and the process players go through as they move through the story helps to reinforce and emphasize the important themes of the narrative.

As a game writer, you may be developing a game on your own and are therefore responsible for both level design and narrative, in which case you will need to think on your own about how to design increasingly difficult game challenges, and how to intertwine story and mechanics so that each strengthens the other. More commonly, though, you will be working as a part of a team, and you'll want to make sure you understand the components the rest of the team is developing, and to consider how you might use gameplay mechanics to highlight and influence elements of the narrative you're creating so that both support one another.

Failure and replay are further important elements to consider when writing game-based stories. In *Sayonara Wild Hearts*, for instance, if a player runs into a tree or gets shot by a villain, the screen freezes for a moment before the player is returned to an earlier point in the level. Similarly, in *Life Is Strange,* if Max fails at a particular task (preventing Chloe from being hit by a train, for instance), the level restarts at an earlier point. Other games may expect players to be responsible for saving their game regularly, or some may take an even more extreme approach and incorporate lasting penalties or require players to start over completely. In Red Hook Studios' *Darkest Dungeon* (2016), for instance, when a character dies, that character is gone for good, and in Klei Entertainment's *Don't Starve* (2013), there are a few special items that allow players to resurrect, but once those are exhausted, they must start a new game from the beginning.

Of course, even if a player is forced to restart a game, they are still bringing lessons and skills they've developed from previous attempts, so it is never a true restart, but as writers, it's useful to think about how failure and replay might influence the stories we're telling. For instance, in *Life Is Strange*, by the time the train encounter happens, Max and Chloe have developed a close friendship, which helps to invest players in saving her, no matter how challenging the gameplay might be in that specific moment. Story also helps pull players through moments of the game that might feel more tedious (such as collecting bottles or escaping a maze), and the potential to develop relationships differently or to make alternate choices encourages players to return to a game even after they've completed it. Because it's difficult to play the same game in exactly the same way twice, game writing might mean creating many narratives that are quite different from one another while also being careful not to prioritize any particular version of the story over any other.

Considering replay and failure can seem particularly daunting as a game writer, because often it means that you're not just writing one story but several, and it is impossible to predict how players will experience the narrative. It's useful to begin to think more about how failure and replay might enhance the stories you tell. What kinds of narratives lend themselves to multiple endings and different outcomes? How might you encourage players to return to a story they've already completed or to keep trying a difficult challenge they have failed repeatedly already? What different outcomes or experiences might be possible within the world of your story, and what can it mean for a story when there is no one "true" ending or experience? Failure and replay are crucial elements of most game experiences, and if, as writers, we can use our stories to help engage with this, the stronger our narratives will be.

Overview

As we've described in this chapter, successful game design considers a number of important concepts such as the magic circle, rules and constraints, choice and actions, resources and tools, signs and symbols, failure and replay, and more. Though it might be tempting as writers to dismiss the principles of basic game design (after all, we're just the writers!), game-based storytelling and design mechanics are closely intertwined. The more we understand these principles and the more we are able to interweave them with the stories we create, the closer we'll be to crafting unforgettable game-based stories that have players returning again and again.

Exercise

(1) Choose one of your favorite games, and consider its elements. What do you notice about how it is designed? What kinds of resources does this game include, and how are they incorporated into the narrative? What themes and symbols are part of this game? How does the game create opportunities for meaningful choice? What do you notice about the magic circle in this game? How does this game encourage players to engage with ideas that might be challenging? How does the game approach failure and replay? In what ways do the game mechanics work with narrative and help to support it?

(2) Then, choose a few mechanics and imagine what would happen to the game if you changed them. How would this game be different if you removed a resource or added an additional one? If this is a single-player game, how could you transform it into a multiplayer one? If the game is competitive, what would it look like if you tried to make it cooperative?

9

Approaches to Game Construction

The Challenges of Storytelling in Games

At the beginning of Inkle's 2014 game *80 Days*, players step into the role of the character Passepartout, a valet tasked with helping his employer, Phileas Fogg, win a bet to circumnavigate the globe in eighty days. A steampunk reimagining of Jules Verne's 1872 novel *Around the World in Eighty Days*, the game *80 Days* is a massive text-based adventure, written mainly by Meg Jayanth. Meticulously researched and carefully designed, *80 Days* allows players to explore an extensive reimagining of the nineteenth-century world, which is packed with technical wonders such as mechanical dragons, flying machines, and artificers, and has been carefully designed to push back against the white, male, colonial perspectives of Verne's original novel.

The basic mechanics of the game are fairly simple. Players must take care of Fogg while choosing how and where to travel. For instance, should they take the Trans-Siberian railroad or travel by hydrofoil? Does it make sense to head to Dubrovnik or to Budapest? Each destination and mode of travel, though, lead to unpredictable story experiences that have been designed to draw players in and convince them to choose engaging with the narrative over winning the bet. At the 2015 Game Developers Conference, Jayanth said, "My job as a writer was to tempt players into making bad decisions [because] a bad strategy decision might lead them to a more interesting story." According to Jayanth, "it's the near-misses, the catastrophes, the daring escapes that players remember and talk about. It's the adventures they weren't expecting to have that make a game memorable."

When considered from a writer's perspective, *80 Days* is something of a marvel. Not only is it massive (over 500,000 words), but the world has been carefully designed with a mind to fostering diversity and inclusivity while also relying on meticulous historical research of the globe at the time. For instance, as a way of pushing back against the Eurocentrism present in Verne's original novel, Jayanth and her team decided that Haiti, rather than North America, would be the emerging economic and political force in the globe in *80 Days*, and as they developed the world, introducing technical marvels and political upheavals, they thought very carefully about how a change in one area (using Zulu technology to avert the Scramble for Africa, for instance) might influence life in another. A playthrough of *80 Days* resembles an actual travel experience in its unpredictability. No matter what Passepartout has planned, it is difficult to foresee exactly what will happen or how a journey will play out, and random game elements make it almost impossible to experience the story in exactly the same way twice. Regardless of where you go and what happens, however, this world is full of fascinating adventures and interesting characters, all of which fit together into a satisfying whole, no matter what path you take.

As writers, an examination of *80 Days* can make creating a game-based story seem very daunting (500,000 words? That's so many!). Any writer who has dabbled with Twine to create a choice-based text adventure realizes how easy it is for one set of choices to snowball into multiple others, until a story becomes an overwhelming avalanche of branches and endings. It's true that compared to traditional print-based writing, creating choice-based game narratives often requires writers to tell the same story more than once and to imagine a variety of different paths and outcomes. Even so, there are things you can do to make writing a choice-driven story manageable and enjoyable. This chapter will explore important elements like considering player types, crafting avatar experiences, understanding common choice and narrative structures, engaging your audience, and more. You'll be well on your way to creating anything, from a two-minute game to a large-scale masterpiece that rivals *80 Days*.

Approaches to Different Player Types

As writers, discussions of craft often center around the concept of audience. We're used to thinking about a hypothetical "reader," and considering the ways the choices we make as authors might influence how someone experiences a certain story. These discussions of audience might take the form of genre considerations. What might a science fiction reader expect from a story experience? What about a reader of romance novels? How can we make sure to surprise these readers while still fulfilling the expectations of the form? This can also take the shape of discussions about demographics and tastes. Writing a story for a young adult audience, for instance, requires certain considerations and approaches that are different from writing for adults. The workshop model utilized in many creative writing classrooms prioritizes consideration of audience by encouraging writers to listen to and think carefully about the feedback they receive from other readers. To some extent, writing for games is no different. Playtesting emphasizes the importance of receiving feedback frequently throughout the process of designing a game and encourages developers to refine and adapt their games as they observe how players interact with them.

A traditional understanding of audience from a creative writing perspective will also serve you well in game writing. People playing science fiction games, for instance, have many similar expectations to those reading science fiction novels or short stories. Because games tend to be more interactive than other genres, though, there is a wider variety of audience approaches to them. Most games allow for a diversity of experiences and playstyles, and so considerations of audience can be a bit more complex. Understanding the different reasons why someone might play a game, and the various kinds of experiences players might be looking for, will help you be even more effective in appreciating and meeting audience needs when writing games.

One of the first and most famous discussions of player types was written by Richard Bartle in 1996, who classified the reasons why players engage with multiuser dungeons (MUDs)—online, multiplayer, text-based dungeon crawler adventure games popular at the time. Bartle's taxonomy categorizes players into four main groups: achievers (who value things like points, levels, loot, and other concrete measures of success); explorers (who enjoy immersing themselves in the world and finding its undiscovered secrets); socializers (who appreciate games for the opportunities they provide for interacting with others); and killers (who value the competitive elements of games). Though Bartle's MUD-based categorization might seem a little dated by today's standards, it helps illustrate the variety of reasons why players engage with games. More recently, Quantic Foundry, a market research company dedicated to understanding gamer motivation, has published a similar series of categorizations intended to help game designers understand their audiences. Quantic's categories include action (players who value destruction and excitement), social (players focused on competition and community), mastery (players interested in challenge and strategy), achievement (players who appreciate completion and power), immersion (players motivated by fantasy and story), and creativity (players engaged with design and discovery). Though each of these measures differs slightly, and though many players don't fit easily into any one category, these classifications help illustrate the various reasons why different people might enjoy playing the same game.

As game writers, these taxonomies may be useful to us in several ways. First, it can be helpful to understand our own motivations for playing games and to understand the types of game experiences we might privilege in our own works, often without realizing it. Both Bartle and Quantic have assessments that can be taken for free online, and that will give you a detailed profile of the things you enjoy most as a gamer. While this may help you make more informed choices about the games you play, it will also provide insight into your own strengths and weaknesses as a writer by giving you a sense of the kinds of experiences you might prioritize. Further, understanding different player types and motivations can help you consider the different reasons why players might engage with a game and can help you make design choices that might appeal to a particular type of player. Trying to appeal to all player types in a given game might be very difficult, particularly for smaller projects, but it is possible to adapt your game to include elements that might appeal to multiple types of players.

For instance, as a text-heavy, story-based game, *80 Days* is probably best suited for players who are interested in fantasy and story. Its massive scope and developed worldbuilding make it appealing to players who value exploration and uncovering secrets, and its open-ended storytelling is best appreciated by creative players who enjoy influencing the game. Though the game may not immediately seem fast-paced or action-oriented, focusing the story around winning a bet helps to incorporate competitive elements as players try to finish more quickly than other players. The game also offers achievement badges, which can incentivize accomplishment-oriented players to replay the game. Further, while *80 Days* doesn't offer much in the way of interaction with other players, it does incorporate multiplayer icons that are sometimes visible. These icons show where other players are traveling on the map, what day they are on in their voyage, and what mode of transport they are using, all of which might appeal to competitive players or those interested in socializing with others. Though *80 Days* may appeal to certain types of players more than others, it's easy to see how the game has incorporated design choices that might engage a broad array of player types.

As a game writer, it's useful to consider how you might adapt your games to make them more appealing to a broad range of players. Many readers of this book will likely prioritize story, and we'll talk more in the chapter about crafting meaningful choices and engaging structures. Are there ways you can encourage exploration in your game, perhaps by incorporating secret areas or hidden secrets? Can you include achievements or other ways of encouraging players to replay your game or to engage with it in unusual ways? What mechanisms might be useful for competitive players or those interested in social elements? Are there strategic elements that could play a stronger role in your game? It's not necessary to try to meet the needs of *every* player in every single game you create, but understanding the different player motivations can go a long way in helping to make your projects appeal to a wide variety of players. Considering player identity is also useful if you're hoping to target a very specific audience with your game. For instance, if you are making a game for an underrepresented community (a game like *Butterfly Soup*, for example), you'll want to consider the needs of that group and perhaps even create a playtest specifically for that community to tailor your game to meet those expectations.

Avatars and Perspectives

One of the most important choices you'll have to make in designing your digital game is the perspective that you'll ask players to adopt while within your story. This is a complex subject, and we talk about elements connected to avatars and character development elsewhere in the book as well. In this chapter, though, we'll be thinking about character development specifically as it relates to story structure and design. How can you make your POV choice work best for your story? And how can you encourage players to connect with your central character, particularly if it is someone who is very different from themselves?

The first question you'll need to answer is whether you'd like players to be able to craft their own avatar identity, either by selecting specific traits in a character-creation screen or by centering your narrative around an indeterminate perspective, often "you" or "I," which fits any player regardless of demographic. Choice of Games, for instance, a developer known for creating a variety of text-based, choice-centered narratives, requires that all player characters in its series be highly customizable. This means that players should be able to select from a variety of options that encompass a diverse array of gender identities, sexual orientations, races, physical abilities, and other demographic elements. If players are given romantic opportunities within the text, Choice of Games requires that they be able to customize those narrative experiences to accommodate a variety of genders, pronouns, and orientations. The company has similar expectations for non-player characters within its titles, requiring a diverse cast of characters in all of its narratives. For them, this ensures that all players feel welcome and represented, and that everyone can interact with or create characters who resemble them in the game.

From a writer's perspective, this level of adaptability may seem daunting. (How is it possible to represent *all* possible demographic experiences in a story?) But it is a crucial element of successfully engaging with customization. The more options you offer for customization, the more likely it is that a diverse array of players will be able to feel at home in your game. Many games choose to center their narratives around an indeterminate first- or second-person perspective, with the idea that allowing the central character to be as unmarked as possible makes it easy for players to connect to the central character without requiring specific customization. For instance, earlier in the book we talked about 3 Minute Games' 2017 title *Lifeline*, which asks players to assume that Taylor, a graduate student who has found themselves marooned on a strange planet in space, has managed to connect with them through their mobile device. Players offer advice to Taylor, via a series of text messages that take place in real time, as Taylor navigates the dangers of the planet. *Lifeline* is careful not to assume specifics about the player such as gender, pronouns, or sexual orientation, and further, the game also avoids assigning a specific gender to Taylor, who becomes a kind of stand-in for players as they navigate the world. Despite these careful choices, however, the game illustrates one of the common challenges of crafting a narrative around a nondescript "you" or "I" perspective. Because the player's responses to Taylor are prewritten choices, it can be easy for a player to feel disconnected from certain statements or put into a situation where they are asked to select responses that don't reflect how they would actually act. Further, choosing not to assign determining characteristics is a choice in itself, in that it allows players to make assumptions about character details, often without noticing it. Indeed, many *Lifeline* players unconsciously assign a gender identity to Taylor without even realizing they have done so.

A third option is to focus your narrative on a specific central character, asking players to identify with a particular perspective and to make choices on this person's behalf. In Campo Santo's 2016 game *Firewatch*, for instance, players are asked to take on the role of Henry, a middle-aged man who has taken a job at a fire observation tower in a park in Wyoming to cope with struggles connected to his wife's early-onset dementia. Henry's circumstances are clearly very different from things most players have experienced, and the biggest challenge facing designers working with this kind of perspective is getting players to connect to and empathize with the central character. *Firewatch* strives to overcome this in several ways. First, in a short text-centered, choice-based segment at the beginning of the game, players are asked to step into the role of Henry from the perspective of "you," and to make choices about his wife and marriage that influence the rest of the game. This asks players to identify with Henry in difficult scenarios right from the beginning and gives them agency in helping to shape his character. Further, since Henry is just starting his new job, players learn about his role and the park along with him, and they participate in tasks like cleaning up discarded beer cans and confronting inconsiderate campers, which helps to foster connection with the character. The game also requires players to navigate beautiful natural scenery, a quiet and peaceful experience which encourages empathy and connection. The developers have consistently made design choices to encourage players to connect to Henry and become invested in his story.

It may seem strange that in a game like *80 Days*, with an emphasis on diversity and inclusion, designers would choose for players to take the role of Passepartout, a white male protagonist. Partly this is due to the game's source material, as Passepartout is the central character in Verne's original novel. A closer examination, though, reveals more nuanced design decisions at work. First, as Passepartout is very close to a default or unmarked state, this gives players the ability to craft him in ways that are most interesting to them by allowing him to pursue both female and male love interests, for instance, or to adopt certain personality traits. Further, as Jayanth has described in multiple interviews, *80 Days* is meant to challenge the idea of a story centering around two white men navigating the globe. Jayanth writes,

> [*80 Days*] is a world where the protagonist's story of racing around the world isn't necessarily the most important, or the most interesting one available to the player. This begs a larger question: is it possible to write a game in which your *protagonist* isn't the hero? Or maybe, less provocatively: can you write a game in which your protagonist isn't the *only* hero?

From this perspective, choosing to ask players to step into the role of Passepartout helps designers to achieve some of the most important goals of the game. For one thing, it invites players to consider their own relationship to the colonial Eurocentric perspectives present in Verne's novel. For another, it encourages players to immerse themselves in the stories of the diverse characters they encounter, stories that are more interesting than Passepartout's and that don't revolve around Fogg's bet. It's clear that the designers of *80 Days* thought very carefully when choosing the POV and perspectives most useful for meeting their goals.

As a writer, when choosing whether to create a game that allows players customization or asks them to empathize with and step into the role of a developed character, you'll want to ask yourself a few questions. First, to what extent is your story character-driven? Is it reliant on a specific perspective or on tracking the change and development of a particular person? If so, a third-person perspective may work best, and you'll want to make sure to incorporate strategies designed to foster empathy and create connections between players and your character. Is your game more player-focused? In other words, would you like for players to consider how they might react in a particular situation, or to encourage them to engage with a story that is closely connected with a specific world or location? Or, is your story intended to draw awareness to some element of that player's own lives? (For instance, the narrative impact of *Lifeline* draws awareness to the close relationships many of us have with our mobile phones.) If this is the case, an undetermined central perspective may work best, one that allows players to participate in the story as themselves. If you go this route, you'll just want to make sure that your game truly is inclusive, that it allows for customization when possible and doesn't fall victim to making unstated assumptions about who your players are and how they will act, particularly in ways that might promote harmful stereotypes or seem non-inclusive. For instance, one critique that students often have when playing *Lifeline* is that the choices offered to respond to Taylor don't always include options that reflect what they would most like to say. This pulls them out of the narrative, as the fictional "you" does not accurately represent them. Similarly, non-inclusive choices involving personal topics like romance, sexual orientation, or gender identity could not only disrupt the illusion of the story but could be triggering or harmful. Whichever route you choose, just be aware that this decision will require thoughtful design approaches and narrative strategies which will influence your game overall.

Structures and Approaches

Once you've decided on which perspective will work best for your game, it's useful to spend some time planning the overarching structure for your project and thinking about what approaches will help make your narrative as strong as possible. We've touched on some common approaches to plot earlier in the book, and, here, we'll talk more about some other common strategies for structuring games. One of the most common problems that new game writers encounter when beginning a project is that it's easy for choice-based narratives to expand quickly into massive stories with overwhelming scope. Beyond being challenging to manage, this can also make it difficult to shape the narrative experience. How can you craft a story that allows player choice while still creating a strong narrative? Luckily, there are some strategies that will help you incorporate choice and give your players a sense of agency while still being able to retain some control over your story.

One of the most common game-based narrative structures is to connect the story closely to a map or environment. For instance, in the game *Stray*, players take on the role of a cat who has accidentally fallen into the ruins of a postapocalyptic world now inhabited only by the robots left behind. Each level of the game corresponds with a different area of the map, and the further a player progresses in the game, the more they learn about the strange world they've entered. Within each region, players are often free to explore in whatever order they like. In the slums, you can explore the rooftops or visit the bar and the shops, but you won't be able to move on to the next part of the story until you've solved the required puzzles, which opens up a new area of the map. Most of the solutions to the puzzles are only possible if players have explored thoroughly, which encourages them to look closely at the environment, thereby learning more about the world, its inhabitants, and what has happened to them. The map approach is common in many games—*Firewatch*, *Life Is Strange*, and *Unpacking* are just a few recent examples—and while it gives players freedom to roam and explore, this structure also often funnels participants to certain key story moments, which everyone must experience if they want to move forward in the game.

Another benefit of linking story to exploration is that you can use the environment to heighten suspense, create tension, or mirror the emotional impact of the narrative. As we discussed earlier, Fullbright Company's 2013 *Gone Home* has become problematic due to allegations raised recently about cofounder Steve Gaynor's verbal harassment of women and his role in perpetuating a toxic work culture for his female employees. Despite Gaynor's deplorable behavior, *Gone Home* is notable as one of the first games to center and amplify the voices of women and members of the LGBTQ+ community, and several women, including Karla Zimonja, a cofounder of Fullbright, also worked on the game, which also features music by feminist riot grrrl bands. It is one of the first and most well-known examples of the walking simulator genre, was widely celebrated by LGBTQ+ players, some of whom saw their own experiences reflected in the game, and is one of the best examples of spatial and environmental storytelling, which is why we have chosen to use it as an example here.

In *Gone Home*, players take on the role of Katie, a young adult who has returned home from overseas travel to find her family's mansion empty, and her parents and sister vanished. The game relies heavily on horror tropes, and the lights flicker and thunder crashes as players explore the mansion to try to figure out what has happened to Katie's family. Much of *Gone Home*'s story relies on map exploration, as players must examine the house, discovering clues to the secrets of each of Katie's family members, and learning more about Katie's sister Sam as she comes to terms with her sexuality and explores her romantic relationship with her female friend Lonnie. Players must solve puzzles to move on to each new area of the house, a design approach that enables creators to allow (somewhat) open player exploration while also maintaining control of the order in which players experience the story.

Gone Home goes one step further, though, in using its map to emphasize and highlight the emotional elements of the game. Near the middle of the game, for instance, players discover the door to the attic, encircled with red Christmas lights, and they learn, too, that this hideout is one of Sam's favorite places. It's clear that this room likely holds the answers, but players are unable to enter it until the end. As they move through the home, they pass under this doorway again and again, and are constantly reminded of the anxiety and mystery of Sam's disappearance, which is amplified by an increasing understanding of Sam's sexuality and the cultural knowledge that for many LGBTQ+ teens, particularly in the 1990s, such identities, and the family conflicts that often arose alongside them, are closely linked to mental health issues like depression and suicide. The climax of the story and the answer to the mystery is literally at the top of the house, and players must climb to it, even as the game's horror tropes foster an atmosphere of fear and reluctance, and a very real worry of uncovering a horrible tragedy.

The reason that Sam might struggle to come out in this family becomes clear in one of the most traumatic revelations of the game, deep in the basement, in the darkest room, which is located spatially almost directly beneath the attic. Near this room, players learn that Katie and Sam's father was sexually assaulted as a boy by his uncle, here in this house, and he has never spoken of it to his family. As players learn about this event, they must cope with their own fear and anxiety, recognizing that the event has happened *right here*, in a dimly lit room in the depths of the house, a physical embodiment of a deep family trauma, which sits at the heart of the house, and also helps to explain the loneliness and distance its residents feel. Music, art, and game design work together to make the basement room both simple and terrifying, emphasizing the emotional impact of that space. These techniques also help to highlight the family trauma and the tendency for families to repress traumas like these, even as they tear them apart. Katie's family doesn't want to enter that room or to acknowledge it, and the players don't either. (The first time Juli played *Gone Home*, for instance, she was too terrified to even bring herself to enter that room.) The basement room's spatial connection to the attic room also helps to explain the revelation at the end of the game that Sam has run away with Lonnie. It would be impossible for Sam to exist as her most authentic self in a home with such a dark room at its center. At the same time, the game offers a sense of hope that Katie's explorations will help to save this family, and that her uncovering of the family secrets will have the effect of turning on a light, eliminating the horror once and for all. Techniques like these use map design, exploration, and gameplay mechanics to create an emotive response in the player that amplifies the emotional tone of the narrative, and they are some of the most powerful tools at your disposal as a game writer.

Using maps or other devices to funnel players through certain choke points is a common structural approach in narrative games, but one downside of this is that it can make subsequent playthroughs seem repetitive. Once a player has learned the answers to the mysteries of *Gone Home*, for instance, what motivates them to play the game again? Another option, though a slightly more time-intensive one, is to allow players more freedom to choose multiple paths through the story, while working to incorporate storytelling techniques like rising tension and an effective climax into each possible path. This is the route that Jayanth and her team took when crafting *80 Days*, a process that Jayanth describes in her 2015 talk for the Game Developers Conference as "building worlds, not plot." Each journey through *80 Days* is nonlinear and completely driven by player choice. Jayanth describes the narrative strategy of *80 Days* as a story told through "accretion," so that each player's journey is a highly different compilation of numerous memorable encounters. Because it is impossible to predict where players will travel, narrative structural devices are tied to regions rather than specific locations. For instance, since players begin their journeys in Europe, the various nodes in that region provide introductory context and information about the world. Once players reach the Americas and West Africa, most are nearing the end of their journey, and so those encounters tend to be riskier, with higher consequences, as a way of heightening tension and suspense. While the structural approach of *80 Days* still draws upon the rising action, climax, falling action pattern common in many print and film narratives, each playthrough follows that pattern in vastly different ways, fostering replayability and making each path through the game highly personalized.

Story structures in traditional print fiction are often closely tied to character, and it is often expected that the main character in a story will undergo some change or transformation from the beginning of the story to the end. If your game asks players to identify with and make choices on behalf of a specific character—Max in *Life Is Strange* or Henry in *Firewatch*, for instance—the structural choices in your game will also need to account for character development and transformation, and you will want to think closely about the extent to which you'd like players to be able to shape your character's personality. In this kind of story, you'll want to consider the important moments in this character's experience and how player choices might lead this specific character down a particular path. In *Life Is Strange*, for instance, Max is put into the position of saving a friend who is attempting suicide, and the story proceeds regardless of whether or not Max succeeds. After this moment in the game, the events that Max experiences are largely the same, but her character is deeply changed by this encounter with her friend, and the story events are colored to reflect that. In *Firewatch*, players develop Henry's character by choosing which responses he gives over his walkie-talkie to his colleague Delilah. The events Henry experiences in the story are largely the same regardless of what players choose, but the way he responds to them is heavily influenced by these dialogue selections. In these two examples, the agency that players experience is closely linked to character development, and participants may be encouraged to replay the game to develop the character differently. In *Firewatch*, for instance, it is possible to foster a close rapport with Delilah or to avoid responding to her entirely, leading to a broad spectrum of possible narrative experiences.

Whatever type of game you are creating, it is useful to spend some time considering structure before you begin writing. Would you like your story to follow a specific trajectory or narrative path, or would you prefer for players to be able to experience a wide variety of potential stories, even if it means you have less narrative control? To what extent would you like for your story to be tied to environment, and how might you use maps or worldbuilding to establish structure? Is this story intended to develop a specific character or are you hoping to draw attention to the player's own choices and implicate them in the story? Would you like for players to be able to shape the central character in your piece, and if so, what kinds of character development might be possible in the story? What moments are most important, and how might you allow players to have a feeling of agency and influence in those key scenes? Digital storytelling allows for a wider array of structures and narrative possibilities than traditional media, so don't be afraid to experiment. Take some time, too, to consider the narrative structures of your own favorite games. How are those stories put together? What kinds of choice and agency are possible? The more time you spend planning how you might structure your game-based story, the less likely it is that you will find yourself overwhelmed by snowballing choices and potential story lines.

Crafting Meaningful Choice

Once you've decided on a structural approach for your game, it's a good idea to think about the role that you'd like meaningful choice to play in your project and to craft a strategy for approaching the moments where player agency will have the highest impact. Choice can be meaningful in a number of different ways, so the kinds of choices you incorporate into your game will depend heavily on the kind of experience you are trying to create. In *Stray*, for instance, players have limited control over the narrative, but they do have control over how they explore the city, and they are responsible for solving the puzzles that allow them to proceed. In *Life Is Strange*, choice is largely based around interpersonal relationships and the decisions players make as they help Max navigate a variety of relationships. In this example, meaning comes from the kinds of relationships and interactions players help Max to foster.

In a game like Lucas Pope's *Papers, Please* (2014), players are asked to imagine they are workers at a passport control booth at a border in the fictional country of Arstotzka, which resembles a cold-war era communist country. In this game, players inhabit a specific world and role, but otherwise they play as themselves, and the story asks them to make moral and ethical decisions. Will they accept a bribe, for instance, or will they allow refugees with incomplete documents into the country to escape violence, even if it will result in docked pay and possible repercussions for their own struggling family? Many of these impossible decisions are meaningful because they turn the camera back on the player, asking them to consider why they have made these choices and the kinds of pressures that might lead someone to make unethical choices in a similar environment. As you're planning your own approach, you'll want to think about the kind of impact you'd like your game to have. Do you want players to become immersed in exploring a gripping story? To develop a complex character? To consider their own actions and choices? Remember, too, that different player types will have different expectations for your game, and that their ideas of what constitutes "meaningful choice" may be very different. You don't need to try to please everyone, but it's a good idea to have a sense of the kind of story experience you want to create, and the choices that will be most meaningful within that context, as well as the kinds of players that you expect will be most engaged by your game.

Before you begin writing your game, it's useful to develop a list of some of the most important choices you'll ask players to make in the game and to consider strategies for giving those key moments weight and importance. The strategy you use for creating meaningful choice in your game will vary depending on the outcome you're hoping for, but here are some useful guidelines to keep in mind as you work to craft choices that are meaningful and that have an impact on the narrative.

The first element to consider is *awareness*. For players to feel that they are making a meaningful choice, they need to be aware that they are choosing something, and it's also important to give them a sense of the context that surrounds that choice. There's nothing more frustrating than making a choice haphazardly and realizing too late that it has determined the whole course of the game, or worse, dying unexpectedly and having to start the game again. Imagine for a moment that you're creating an environmental exploration game where players are given a choice of two doors. In a simple version, you might include two choices, "go left" or "go right," where one door leads to immediate death and the other allows players to continue the story. If you don't give players clues that one door might lead to instant death, or information which allows them to make a thoughtful choice between them, this is likely to be an unsatisfying narrative experience. Creating awareness would mean giving players hints as to what choosing one door over the other might mean. One door might have a carving of an ocean on it, for instance, and the other an image of a meadow, or you might incorporate sensory environmental details such as smell or texture. If a choice has immense consequences, it's also useful to foreshadow it, giving players ample time to gather information and to consider what some possible outcomes might be. In the previous example, perhaps players have encountered a legend about these two doors earlier in the game, and they are aware that each leads to separate regions of the world. Further, they are aware that once they have chosen one door, they will be unable to go back and choose another.

This leads us to the second element that is important to creating meaningful choice, which is *consequences*. For a choice to feel meaningful, there should be a definite outcome that is clearly linked to the player's choice. Players need to be aware of how the choice they've made has influenced or affected the game. In *Papers, Please*, accepting a bribe might lead to better living conditions temporarily, but it shouldn't surprise players later when they are caught and taken to jail, ending their game and forcing them to restart. Consequences can sometimes be fatal, as in this example, but often choice is more meaningful if either option leads to a viable story path, so that players feel they have shaped the story rather than simply failed at it. In *80 Days*, for instance, Jayanth and her team wanted players to feel encouraged to make "bad" choices for the sake of a more interesting narrative, and it is very difficult to die in the game and to be forced to restart. Instead, choices have narrative consequences. For instance, it's possible for players to choose to travel on a slaver ship, returning to Africa after delivering its cargo, but once they arrive, they are mistaken for a slaver and treated as such, a character development which is memorable and meaningful.

The third element to consider when crafting meaningful choice is *impact*. For a choice to be meaningful, it's helpful for players to be reminded of how this decision has influenced the story. In the previous *80 Days* example, for instance, the choice to travel on a slaver ship would feel much less meaningful if players weren't reminded of that decision in conversations with residents afterwards, or if the story didn't require them to transport a slave in the next leg of their journey. In *Life Is Strange*, Max thinks of the friend who struggled with suicide several times after the event, regardless of its outcome, reflecting on this encounter and the impact it had on her. It can also be useful to give players quieter moments in the narrative so that they have time to process the emotional weight of their choices and to reflect on what has happened. Whatever approach you use, it's up to you as a writer to help players understand the meaning that their choices have had on the trajectory of the story and the characters they are helping to shape.

Permanence is the final element to consider when crafting meaningful choice. In order for a decision to feel important, players need to understand the choice isn't reversible and that it has lasting consequences. In *Life Is Strange*, for instance, players are able to rewind time to make small changes to most conversations, but once they've chosen a trajectory, there is no going back. If you accept the bribe in *Papers, Please*, you will need to play through the game again if you decide later that you'd prefer to leave the money on the table. And if you'd prefer to not be associated with slavers in *80 Days*, you'll need to remember the lesson and avoid taking the slaver ship on your next playthrough. As with consequences, permanence is most useful if the story is equally viable regardless of which choice the player has made. Rather than killing the player or ending the story in other ways, permanence will help make a choice feel more meaningful if players are able to continue the game and see the effects that their choices have on the story. Interest in experiencing another outcome might even make them decide that it's worth the time to play through the game again after finishing it.

As with game structure, developing effective and meaningful choices requires a certain amount of planning and development. You'll need to think carefully about the kind of story you are telling, the types of players you're most likely to engage, and the impact you'd like for your narrative to have. A little preparation, though, can go a long way in crafting a story that is memorable and significant and that has players returning to your game again and again.

Case Study: *The Stanley Parable* (2011, 2013, 2022)

Click "Start" on 2011's *The Stanley Parable*, and a narrator says, "This is the story of a man named Stanley." One day, Stanley gets up from his desk and discovers that everyone in his office has disappeared. You wander around the office and the narrator simultaneously narrates your actions *and* the actions you're about to take. For example, the player enters a room with two open doors and before you can make a decision the narrator says, "When Stanley came to a set of two open doors, he entered the door on his left." Follow the narrator's instructions and the game will end rather abruptly. Stanley will discover that his job is connected to a mysterious surveillance system. In the end, he breaks out, and the narrator describes how Stanley has successfully escaped from control and can now become free, ignoring external commands. Before the player can really untangle any of this, the game restarts. Before long, Stanley is again faced with the same set of two open doors, and the narrator repeats, "When Stanley came to a set of two open doors, he entered the door on his left." But what happens if you go right (Figure 11)?

When Stanley came to a set of two open doors, he entered the door on his left.

Figure 11 *The Stanley Parable* dares to ask the fundamental question: What is choice?

Initially, the narrator says, "This was not the way to the meeting room, and Stanley knew it perfectly well." He'll wait for the player to turn around and recommit to the story as conceived, but if the player refuses and barrels headfirst into uncharted territory, *The Stanley Parable* will spiral outward into nineteen different endings, each more bizarre than the last. The narrator will plead with the player at multiple points and will grow increasingly frustrated. If pushed far enough, the narrator will literally remove the player from the game, dropping them into a brief curio of *Minecraft* or *Portal*. Provoked further, and the player will descend into the liminal space beyond the game where they can explore blank rooms lacking textures.

The Stanley Parable is concerned primarily with the question of what a video game story is and how it functions. How do you reconcile the freedom of player choice with the rigid confines of a frame narrative? This tug of war between player and writer—illustrated here between Stanley and the narrator—is explored with meta irony like a *Saturday Night Live* sketch. Earlier in this book, we claimed that all video games present the players with rules. *The Stanley Parable* encourages you to break them.

One of our favorite exercises while teaching *The Stanley Parable* is to run the game live in-class and allow the students to vote on choices. Very quickly, they'll discover just how far you can push the game to its extremes. Perhaps they'll jump out the window into the blank empty sky where the narrator sings a song to pass the time. Perhaps they'll vote to hide in the broom closet long enough for the narrator to scold our indecisive Stanley before forcibly restarting the game. Perhaps they'll push Stanley toward the museum ending in which the player can inspect various artifacts relevant to the development of the game. No game more easily defines the struggle of writing narrative for a medium in which the player has control than *The Stanley Parable*, and Galactic Café—the company responsible for the game—tripled down on this when they released *The Stanley Parable: Ultra Deluxe* in 2022. Here, five new endings are added to the original game's nineteen, and now they grapple with what it means to be a remaster of a decade-old game. *The Stanley Parable* does not try to nail an old media story to the slippery surface of video games. Instead, Galactic Café has dreamt up a story that can only be told through the medium of video games. This makes it an object lesson for anyone interested in the form.

Overview

Writing a game-based narrative that gives players influence over the story and its progression might seem daunting and overwhelming, but for many players the opportunity to participate in the creation of a story is one of the most engaging elements of games. It can be helpful to consider different player types and what kinds of expectations they may have in order to refine the appeal of your game. Structuring your story around exploration of a world or a map can help you maintain control over the story while still allowing for player agency and interactivity. Finally, meaningful choice is a defining factor of gameplay, so learning how to craft meaningful choices and center them within your narrative can help you make the games you create even more memorable and compelling.

Exercises

(1) Choose a game that you know well and consider its structure and approach. What elements of this game's story are most linked to player choice, and how have the writers structured the game to lead to those moments? What choices are the most meaningful, and what helps to make those moments have weight? To what extent is this story centered around the exploration of a map?

Once you've spent time deconstructing the game, choose a few of those structural approaches and try them in a creation of your own.

(2) Draw a map of a location that interests you and that seems compelling. (When Juli teaches this lesson, she likes to use an abandoned amusement park.) Consider how each area leads to the next and be specific about entrances and exits. (Can players move directly between the rickety carousel and the tilting Ferris wheel, for instance, or do they need to pass through the gift shop first?) Once you have about ten locations on your map, select a monster (Juli likes to use werewolves). Place that monster in one of the far locations of the map and then imagine that players will work their way toward this monster. You need to give them clues to the monster's existence, but you won't be able to predict exactly which path players will take, so each area should provide something unique. As players approach

the monster, they'll need to have a sense that they're getting closer even if they don't know exactly what they'll be encountering. Make a note of details you'll use in each section to create a sense of increasing suspense.

Once you've created your map, use Twine or Bitsy or the game program of your choice to turn this map into a game, where each location on the map is its own screen or room. Then conduct a playtest and ask players about their experiences. What clues did they find? Were they prepared for the monster when they encountered it?

10

Representation and Inclusion

Gamergate and What It Means for Writers

In 2013, video game developer Zoë Quinn released *Depression Quest*, a text-based game that puts players in the position of someone with depression, asking them to make choices about everyday life—things like visiting doctors, adopting a pet, and spending time with friends. *Depression Quest* was created to help players understand what it's like to have depression and the difficulties that people struggle with when trying to get better. As a text-based game, *Depression Quest* uses grayed-out choices, which players are unable to select unless they meet certain prerequisites, and a scoring system that keeps track of things like a protagonist's mood and whether they are taking antidepressants. Quinn developed the game to bring awareness to this common mental health experience and to help players understand more about depression and the challenges individuals face when recovering.

Depression Quest was released to largely favorable reviews, and it still stands as an excellent example of how games can be used to draw awareness to complex real-world experiences and bring about change. At the time, though, some players (mostly cisgender, heterosexual white men) felt that *Depression Quest*, a game without flashy graphics and glitzy mechanics, was undeserving of the praise it was receiving. They accused Quinn of exchanging romantic and sexual favors for positive media reviews, which wasn't true. Some of the critics couched these accusations as an attempt to ensure ethical treatment of video games in journalism, accusing the media of unfair and unwarranted positive coverage of games that didn't deserve it—often games that engaged with social issues or focused on underrepresented characters and women, LGBTQ+, or BIPOC players. For many, though, this was just a way of rationalizing widespread discrimination in games and reacting to inclusive changes happening in the industry, as well as policing ideas of which games should be made and celebrated, the types of players welcome in games, and the kinds of characters represented.

The term "Gamergate" came from a popular hashtag that was used to talk about the issue on social media, often in connection with horrific online attacks on high-profile women associated with the games industry such as Quinn, Brianna Wu, Anita Sarkeesian, and Felicia Day. The attacks ranged from offensive memes and games that showed these women being beaten, to online threats of violence and death, and doxing (the public release of private information such as addresses and phone numbers), which meant that many of these women fled their homes in fear. For many underrepresented players and developers, the worst thing about Gamergate was that it wasn't a surprise. It reflected ongoing systemic issues in gaming that had always been present (and are still present) and foregrounded discussions about what kinds of people "should" be permitted to be messengers or ambassadors for digital games. Gamergate highlighted the discomfort that many white male designers and players felt with having women and other underrepresented groups representing the medium, a discomfort intense enough to compel some to heated online harassment and threats of physical violence. Finally, for the first time, these issues had come to the awareness of the broader public, leading to a widespread discussion of representation and inclusivity in gaming that is still very central to the industry today.

As game writers, the considerations connected with Gamergate are vital to the work that we do. As storytellers, it is up to us to ensure that the game worlds we create and the characters within them are diverse and representative, that they reflect a variety of perspectives and experiences, and that they are welcoming to as many different kinds of players as possible. We must also understand our own implicit biases, the systemic pressures within the games industry—such as the inequalities that still exist in many game companies—and the larger social forces that shape ideas about which stories are welcomed and by whom. This is challenging work, with very high stakes, and it requires an ongoing commitment to self-reflection and determination. This chapter will help you examine diversity and inclusion in your own work by giving an overview of important things to consider and suggesting useful resources for further research.

Beginning Considerations

One of the first places to start when developing our approach to diversity and inclusion in our own work is by considering our own implicit biases. In the book *Writing the Other*, Nisi Shawl and Cynthia Ward offer practical advice for writers who are interested in treating diversity in their works ethically and responsibly, and they begin their book by asking readers to consider their own identities. Shawl and Ward use the acronym ROAARS, which stands for Race, (sexual) Orientation, Age, Ability, Religion, and Sex. For them, these categories tend to be the most culturally divisive, contributing to widespread problems such as discrimination and stereotyping. Shawl and Ward's approach highlights the intersectionality of identity and the ways that our experiences and perspectives are shaped by a variety of factors. Even if a person aligns with a privileged American cultural majority in some ways (being white and male, for instance), they may differ from it in other ways, which may be sometimes invisible to others (being Jewish and homosexual, for example). In this way, our lives and experiences are shaped by the intersection of a variety of different identities, and many of us may enjoy cultural privilege in some instances, while being part of an underrepresented minority in others.

Shawl and Ward often write about the biological tendency for the human brain to make swift, broad categorizations and assumptions in order to determine the degree to which a particular situation is safe. This inclination is often immediate, and because it is rooted in survival, it is useful and important. These associations happen very quickly, though, and this part of our brains frequently draws on implicit biases and stereotypes that have been ingrained in us since birth through cultural messages, many of which are harmful. According to Shawl and Ward, this biological tendency makes it likely that many of us will have thoughts that are discriminatory (for instance, upon seeing a heavy person, we might make judgments about that person's health and lifestyle), but this is only the psychological impulse to evaluate situations quickly, and it is up to us to use our rational brains to push back against these implicit biases.

This process is ongoing and eternal, and for writers it is particularly important. For one thing, stories and media are some of the primary vectors for shaping our implicit biases, as we tend to see the same scenarios again and again in films, books, and movies, and these repeated stories shape our mind's split-second evaluations. As writers, when we are caught up in creating a story, we may unconsciously draw on common tropes and stereotypes, perpetuating harmful depictions without even realizing it. Further, if we allow these messages to persist in our work, we are contributing to this cultural environment, disseminating problematic portrayals that will shape others' implicit biases. Because of the pervasiveness of these messages, it is likely that we will make mistakes. This doesn't make us "bad" people, though the experience can be uncomfortable. It is up to all of us, though, to push through our discomfort and anxieties, and to do everything we can to make sure that the stories we tell are responsible and ethical, and that they highlight the diverse perspectives and experiences of our world.

As game writers, being committed to diversity and inclusion is particularly important. Though controversies like Gamergate have led to important conversations and useful changes, the game industry still lags behind others in terms of inclusiveness and representation. While most estimates of player populations tend to find players fairly evenly split along gender lines, male game developers still outnumber female developers in many companies. Though representation has improved in recent years, video game protagonists, too, are often male. These statistics are even more problematic when considering the limited presence of LGBTQ+ and BIPOC individuals within game companies and within games themselves. Further, underrepresented groups are more likely to experience harassment from other players, and they are also more likely to report discrimination and unsafe working environments in the game industry.

As writers, our games can never exist entirely separately from industry injustices, and it's important to bear these things in mind when creating our work. It's common for critics to focus on addressing inequalities in the workplace, claiming that once there are more women, BIPOC, and LGBTQ+

individuals in the industry, better representation in games will logically follow. This argument is problematic, though, on a few levels. For one thing, it puts the responsibility for ethical representation in games on underrepresented people, expecting them to endure harmful working environments and unsafe online experiences, with the questionable promise that this will eventually lead to better representation. For another, it ignores cultural and commercial pressures which are often the rationale for problematic choices (the argument that highly sexualized female characters sell more games, for instance).

As players, it can be difficult to get a sense of the working conditions that women, people of color, and other underrepresented groups face at a particular company, which can make it challenging for people wanting to make ethical choices about which games and companies to support. For instance, many *World of Warcraft* players were stunned in 2021 when the state of California announced its lawsuit against Activision Blizzard King (ABK) for widespread workplace harassment and discrimination faced by women and other underrepresented groups within the company. Players quit the game in droves, determined to withdraw financial support from the company, even though, for many, this meant leaving behind communities of supportive players and close friends. (For instance, this lawsuit decimated Juli's online LGBTQ+ and women-friendly player community, one of her safest and favorite gamer groups.) In the aftermath of the lawsuit, ABK made changes including leadership adjustments, the creation of community focus groups, and developed non-toxicity agreements that players were required to sign. Still, many players wondered, could the company ever do enough to make up for past injustices, and how would players ever really know what working conditions were like for women and people of color within the company?

Aware of the weight that these concerns hold for many players, several companies are trying to be more transparent about issues of diversity and inclusion. Some post public DEI statements on their websites, outlining a commitment to fostering diverse and safe workplaces, representing a variety of people and experiences in their games, and supporting nontoxic, inclusive player communities. Others have tried to be more open about the experiences that underrepresented employees have at their companies. As part of Pride Month and Asian American and Pacific Island Heritage Month, Square Enix, for instance, posted interviews with BIPOC and LGTBQ+ employees on their webpage to give those voices more support and visibility. And many companies have been working to diversify their writing rooms and employee populations, often with visible results. Adding women writers to the *Tomb Raider* franchise, for instance, helped to transform Lara Croft from the sexual object she was in the first game into a more complex, independent, and less objectified protagonist of the later ones. Still, these issues are often intersectional and challenging to navigate. *Tomb Raider* may be a great example of a game featuring a strong female protagonist, but it still features a colonial perspective, one where white, Western characters invade indigenous communities and take their artifacts. Corporate DEI statements and required player non-toxicity agreements are flashy and public-facing, but to what extent do they really shape player or workplace experience, or promote actual, meaningful change?

Clearly, these issues are systemic, massive, and complex, and beyond the power of any one individual to repair. As writers, though, one of the most powerful and useful things we can do to combat injustices like these is to make a commitment to prioritizing ethical representation in the things we create. We may not be able to change the industry on our own, and our implicit biases make it likely we will make mistakes. Even so, ensuring that we consider diversity and inclusion in every piece we create can help to improve representation in the industry and contribute to lasting social change.

Character Representation and Worldbuilding

One of the first places to start when considering representation and inclusivity in your own games is to examine your characters. To what extent does your game include characters of different genders, races, ethnicities, physical abilities, sexual orientations, ages, and life experiences? If your game is set in a time and place much like our own, does it include a diversity of perspectives that largely reflects the variety of our actual world? If you are writing about a speculative world that is quite different from our own, what kinds of perspectives and experiences are possible in this world, and to what extent does your game represent them? As we described in Chapter 9, if your game allows players to craft their own customizable character, does it allow for a variety of options? Do the choices available in your game truly allow players to experience the world from a variety of avatar perspectives? For instance, if there is romance in your game, is it possible for players to choose a variety of sexual orientations? To engage with those story elements from a variety of gender identities? To experience those elements from asexual or aromantic perspectives? Remember, too, that ethical representation extends to non-player characters as well, and you'll want to be sure that the minor characters you include in your game reflect the full diversity of the world in which your story is set. Strong examples of this work include *Stardew Valley*, *Hades*, and *Life Is Strange: True Colors*, which all allow players to play as LGBTQ+ characters among other identities.

As writers, it's common and comfortable to write about perspectives that are similar to our own, and so you may find that you have a tendency to create characters with perspectives that are familiar to you. Examining your game with a mind to diversity may mean literally going through and tallying the demographics of your characters. How does the number of male characters compare to the number of female characters and to the number of characters who fall outside of traditional gender binaries? If your game world includes multiple races and ethnicities, do your characters mirror the diverse proportions of that realm? To what extent do the characters in your piece reflect a cross-section of the world? Once you've looked closely at your characters, you may need to make changes to ensure that your characters reflect a variety of backgrounds and perspectives. Of course, this doesn't always apply to games that primarily or entirely showcase characters or identities that are atypically featured in video games. For example, *Butterfly Soup* primarily features Asian queer women. This narrative decision both speaks to power and the unfortunate tradition of games often defaulting to white male characters as default.

While basic calculations are a great place to start, it's important to remember that representation goes beyond surface level. Changing names and character descriptions are a step in the right direction, but you'll need to go even further to ensure that you're representing diverse perspectives ethically. A character's perspective on an occurrence will be different depending on the position from which they are viewing it. A woman waiting to meet a stranger at a shadowy nightclub, for instance, will experience that event differently than a man, and a genderqueer or transgender person will have a different perspective entirely. Social and historical events may mean different things to different individuals, even if those events happened long ago. A game experience like *Red Dead Redemption 2*, for instance, which romanticizes the golden era of the American West, might mean one thing to a player who is white and male, and something else entirely to a player who is a Native American woman. Writing diverse perspectives will require you to research and learn more about the communities and individuals you are representing, and to listen closely and do your best to understand the nuances of those viewpoints. It is also a good idea to find playtesters from those communities and to ask them for feedback about representation in your games. You'll want to make sure that you compensate players for their time, listen carefully to any criticism you receive while keeping an open mind, and make any necessary revisions to improve your work. Our brains have been programmed by years of messaging from the world around us and so we will all inevitably make mistakes. A great place to start is by prioritizing ethical representation, listening carefully to diverse perspectives, and being willing to make changes and to continue learning.

It's important to remember, too, that avoiding choices about diversity is itself a choice. It may seem tempting to go the opposite direction in your game by making character descriptions as vague as possible, for instance, or avoiding names, pronouns, and demographics entirely. Why not leave it up to players to imagine the specific identity experiences of the characters? Isn't that more empowering and inclusive? In *Writing the Other*, Shawl and Ward refer to characters without specific demographic descriptions as existing in an "unmarked state." Particularly common for minor characters, being "unmarked" means that a writer has neglected to include descriptions that "mark" a character as being of a particular race, gender, age, or other demographic. The problem with leaving a character "unmarked," as Shawl and Ward point out, is that often in the absence of descriptive markers, readers (or players) will assign traits to that character automatically, and often, those traits will mirror the most common and privileged perspectives. Further, readers are often unaware that they have assigned those traits to that character. For instance, if you include a shopkeeper in your game but neglect to describe that character, many players will automatically assume that person reflects an unspoken default, likely imagining them as male, white, physically able, and heterosexual. It only takes a few words to tell your players that the shopkeeper's name is Kamila, that she comes from Puerto Rico, and that her arthritic fingers tremble as she opens the box of ammunition. Doing so helps you retain your power as a writer and honors your commitment to diversity in your work.

Avoiding Stereotypes, Tropes, and Harmful Clichés

Writing diversity effectively often starts with a great deal of research and engagement. In order to represent a perspective accurately, we must first understand it, which requires listening and paying close attention. Imagine that you are writing a game that takes place in contemporary America and you'd like the world of your game to reflect the diversity of our real world. Perhaps, you are not a member of the BIPOC community and you'd like to incorporate several Black characters, or maybe you are cisgender and straight and would like to present LGBTQ+ perspectives, or maybe you're a member of a particular political party and you would like to respectfully represent the perspectives of another. Regardless of your own identity and goals, you'll want to have a developed understanding of perspectives that differ from your own in order to communicate those viewpoints accurately and ethically. The perspectives of members of a demographic, though, are not monolithic and representing them accurately will take research and commitment. Among the Black community, for instance, a character who identifies as African American might have a different perspective than an African immigrant, and the nuances of those viewpoints would depend a great deal on where in America the person lives, what African cultures and ethnicities they identify with, and how those experiences have shaped the character's sense of identity, as well as their desires, fears, and worldviews. Depending on your own background or experience, you may need to research heavily, reading books by authors of these communities, visiting websites and online forums, or reaching out to community members to ask for feedback and advice. This is detailed and challenging work, and it might seem overwhelming and full of potential pitfalls, but it is important to do so, particularly if you occupy a place of privilege and are hoping to help curate safe spaces for others.

It can be easy, particularly when we're writing quickly, to inadvertently perpetuate common stereotypes, clichés, or harmful perspectives, and it is important to take the time to examine our creations carefully and critically, with an eye to representation, and to prioritize locating these problems and fixing them. Learning about problematic tropes and common stereotypes is one of the most useful things we can do as writers trying to engage with ethical representation in our work. The more we know about common problems, the easier it is for us to identify them and adjust them. It's likely that you're already familiar with many common problems and stereotypes. Discussions about problematic representations of race— casting Black characters only as villains, for instance, or including characters of color primarily as victims of violence—have become more central to discussions of games and popular culture. Some series like *Grand Theft Auto* and *Call of Duty* have been criticized for their treatment of diversity—for things like perpetuating stereotypes or treating diverse cultures as monolithic by ignoring differences in beliefs. Others, like the game *Detroit: Become Human*, for instance, have been criticized for appropriating real-life historical events and racial discrimination and representing them without depth, nuance, or context. In contrast, games such as *Life Is Strange* and *The Last of Us II* may be held up as examples of works that engage with diversity successfully—through things like representing racial discrimination and transgender identity thoughtfully and empathetically.

The Last of Us II is a particularly interesting example, because while some players celebrated its commitment to strong female characters and LGBTQ+ relationships, others were angered by the decision to kill off Joel, the first game's white, male protagonist, which sparked heated online discussions and toxic conflicts. The conversations surrounding *The Last of Us II* are a reminder that different player groups may have opposing responses as well as intense emotional investment in the characters and situations in a given game. As much as it may be tempting to believe that we have moved on, as a culture, from Gamergate, these discussions are clearly still very charged and important. As game writers who are also game players, one of the most useful habits we can get into is looking critically at the treatment of things like race, gender, and sexuality in the games that we play. What do we observe about how developers of our favorite games have engaged with diversity? What problems do we notice? How have people from underrepresented communities responded to those games? What can we do in our own games to avoid making similar mistakes?

Luckily, too, there are many resources to turn to when learning to write diversity more effectively. In *Writing the Other*, Shawl and Ward list some common problems to avoid:

- Casting heroes and villains along racial or gender lines or linking underrepresented characters repeatedly to problematic roles such as criminals or victims.
- Romanticizing or fetishizing characters of color, such as the "noble savage" or the "exotic," oversexualized Asian love interest.
- Problematic use of vernacular.
- Including "sidekick" characters of color whose primary role is to support a central white character by helping them through various trials.
- Focusing exclusively on traumas (slavery, for instance) that are associated with an underrepresented group rather than allowing

those characters to have stories focused on positive experiences like acceptance, love, and joy.

Choice of Games, a company focused on publishing inclusive, diverse text-based games, also has an extensive discussion about the importance of representation on their website, including discussions of best practices, common problems, and reasons why writers of choice-based digital stories should prioritize ethical representation and inclusivity in their games. Choice of Games also includes links to a variety of resources geared to helping writers engage responsibly with perspectives other than their own. *Writing with Color* is a blog focused on helping writers to think critically about representations of people of color in their work, including extensive discussion of common problematic racial tropes and stereotypes. The GLAAD website provides guidance for ethical representation of LGBTQ+ perspectives in the media, and the National Center on Disability and Journalism is a great resource for questions about disabilities. These are just a few of the many resources available, and it can be useful to keep a running list of sources such as these to keep up with relevant discussions, new terminology, and important criticisms that will help you engage responsibly with perspectives and experiences that are different from your own in the things you create.

Games and the Potential for Reimagining

It's important to understand common representation issues and to always be on the lookout for potential problems. It's even more useful, though, to understand the transformative potential for games, and to recognize the unique opportunities games offer for communicating underrepresented views and experiences in new and vibrant ways. Beyond being concerned simply with avoiding problematic representation in our creations, it's useful to think about potential. How might games engage with diversity differently than other media? How can we use these tools as storytellers to engage with alternative viewpoints even more deeply or to help players understand experiences in ways that might not otherwise be possible?

When Meg Jayanth set out to retell Jules Verne's novel *Around the World in Eighty Days* in Inkle's 2014 release *80 Days*, one of her most pressing concerns as a writer was how to approach diversity and inclusion in the work. First published in 1872, Verne's novel reflects the white, male, heterosexual, and colonial perspectives of the time, views which, at best, do not reflect the variety of experiences in our contemporary world and, at worst, are sexist, racist, and potentially triggering for modern audiences. Jayanth and her team might simply have chosen to adjust the problematic plot points and story lines, updating them for contemporary players, but instead, they took a much more complex approach. In a 2014 post entitled "Family, History, and Respect" on Jayanth's blog, *You Can Panic Now*, she writes, "We wanted our inventions and devices to be grounded in local cultures rather than overwrite them with a purely British notion of steampunk. I wanted to write an anti-colonial adaptation, and part of that was trying not to appropriate or disrespect people's struggles and history." Rather than inventing an entirely new fantastical world, Jayanth's team instead did extensive historical research, focusing on the actual sociocultural realities of the world at the time of Verne's novel, rooting the steampunk elements of the game in the actual societies, traditions, and histories of the cultures represented in the game. In this way, the fantastical elements of *80 Days* build on and draw from the cultural histories being represented, rather than overwriting them.

Further, Jayanth used fantastical approaches in the game as a way of imagining potential and growth, using speculative writing to give underrepresented voices agency and power rather than reinscribing stories of historical violence and trauma. According to Jayanth, this approach allowed the team

> To tell the kind of story we wanted to be able to tell, to redress some of the colonialism, sexism and racism of the period. If you're inventing a world, why not make it more progressive? Why not have women invent half the technologies, and pilot half the airships? Why not shift the balance of power so that Haiti rather than barely postbellum United States is ascendant in the region? Why not have a strong automaton-using Zulu Federation avert the Scramble for Africa? Why not have characters who play with gender and sexuality without fear of reprisal? History is full of women and people of color, and queer people, and minorities. That part isn't fantasy—the fantastical bit in our game is that they're (often but not always) allowed to have their own stories without being silenced and attacked.

By approaching diversity in *80 Days* in this way, Jayanth and her team aren't simply telling a story ethically and responsibly or working to avoid problematic representation (though they succeed in both of those things).

Instead, they use speculative writing, historical research, and fictional revision to essentially imagine a version of the world that honors the diversity of the time, providing a space for underrepresented voices to exist without fear. In this way, *80 Days* gives players the chance to participate in this reimagined world, to see it through new eyes, and to experience the power and potential of what might be possible.

One of the most compelling reasons for using a game to describe an alternate version of the world as Jayanth's team did in *80 Days* is that players are active participants in the story rather than passive observers. Instead of simply watching a story unfold in a diverse steampunk world or observing another person's struggle with depression, games like *80 Days* and *Depression Quest* expect players to act in these worlds, to make choices and face the consequences of those decisions, which can make those experiences (and the diverse perspectives represented in them) much more personal, immediate, and vibrant. In a game like Ubisoft's 2014 release *Valiant Hearts: The Great War*, for instance, players aren't simply being told about the First World War or watching actors engage with it. Instead, they are navigating the trenches, dealing with explosives and poisonous gases, and striving to treat injured civilians themselves. These successes and failures are stressful and adrenaline-inducing, which makes them memorable and therefore messages about diversity, (*Valiant Hearts*, for instance, works to highlight the oft-overlooked contributions of communities of color) can be very powerful. Games, with their emphasis on immersion and interactivity, provide a level of engagement with diverse perspectives that may differ from the player's own and can be particularly strong, arguably more compelling than any other medium (which is perhaps one reason why debates about representation in games are particularly heated).

Further, games offer rhetorical tools that can allow storytellers to create narratives that ask players to engage with diverse messages and experiences on an entirely different level than in other media. As an example, consider E-Line Media's 2014 game *Never Alone* (*Kisima Inηitchuηa*), which was created in partnership with the Cook Island Tribal Council as a way of helping audiences to understand traditional Iñupiat culture and values. Designed in hopes of encouraging younger tribal members to learn about their disappearing culture, *Never Alone* follows the Iñupiat girl Nuna and her arctic fox companion as they travel through a challenging winter landscape and encounter characters from Iñupiat folklore, such as the Little People, Blizzard Man, and the Sky People. On the surface, *Never Alone* is a platform adventure, where players switch between Nuna and the fox (or work cooperatively in a two-player game) as they navigate the world. As players complete levels, they unlock a series of videos where tribal elders explain important elements of Iñupiat culture and folklore, which correlate to the levels and stories Nuna and the fox are experiencing in the game.

In addition to providing information about Iñupiat culture, however, utilizing the game format to present these stories enables *Never Alone* to go even further. As players learn important lessons about cultural values and belief systems, they must use those concepts in order to proceed in the game, in essence enacting these cultural values and participating in them. For instance, through the game and the videos, players learn the Iñupiat belief about the ways the spirit world is interconnected with the real one. Then, they must use spirit helpers to make it through the levels, and play as the fox, who is itself a spirit, thereby taking on the role of a spirit helper themselves. Players learn about the Iñupiat reliance on nature, but then must go even further to pay close attention to natural forces in the game, through mechanics associated with wind, ice, trees, and wildlife in order to progress through the levels. In this way, game design and interactivity work together with narrative as an important rhetorical tool that not only communicates important cultural messages but trains players to participate in them, encouraging deep and nuanced understandings.

As you can see, approaching diversity and inclusion in game-based storytelling goes far beyond simply considering representation and avoiding clichés and stereotypes. When done thoughtfully and well, storytellers can use the principles of game design to craft vibrant, interactive messages and to convey underrepresented viewpoints and perspectives in revolutionary new ways, techniques that have the potential to engage players on a variety of levels. Even more importantly, learning to represent a variety of perspectives and experiences that differ from our own makes us stronger and more versatile storytellers. It's unfortunate that Gamergate and other problematic events have contributed to the stereotype of games and game communities as toxic realms that foster discrimination, bullying, and harmful tropes, when in fact there is great potential for games to work as innovative mechanisms for communicating diverse perspectives in revolutionary new ways.

Overview

As a game writer, it's important to understand the complex systemic injustices that have shaped the game industry and continue to affect player communities today. The works we create can never exist entirely separately from these issues, and as storytellers, we play a great role in influencing the views and underlying belief systems of our audiences. It is crucial that we work continuously to investigate, question, and reform our implicit belief systems, and to demand ethical representation from the games we play and the companies that create them. These systems are long-standing and massive, but we can each work individually to avoid cultural stereotypes, problematic sexualization, and offensive messages in our works, and to ensure the stories we create are as diverse, inclusive, and welcoming as possible. Further, by understanding the unique affordances of game design and game-based storytelling, we may find it possible to create diverse narrative experiences that are more revolutionary, compelling, and inclusive than ever before.

Responsible representation comes through our own ongoing efforts to learn and grow through a commitment to research, to listening to and compensating diverse communities of playtesters, and to being fearless about confronting our own prejudices and internalized belief systems. This field is constantly developing, and we encourage you to research widely and often. The following texts are a good place to start: *Writing the Other* by Nisi Shawl and Cynthia Ward, *Craft in the Real World* by Matt Salesses, *The Anti-Racist Writing Workshop* by Felicia Rose Chavez, *Toward an Inclusive Creative Writing* by Janelle Adsit, *Gaming at the Edge: Sexuality and Gender at the Margins of Gamer Culture* by Adrienne Shaw, *Rise of the Videogame Zinesters* by Anna Anthropy, *Cooperative Gaming: Diversity in the Games Industry and How to Cultivate Inclusion* by Alayna Cole and Jessica Zammit, and *Gamer Trouble* by Amanda Phillips.

Exercises

(1) Choose a game that you've enjoyed playing and spend some time considering it from the perspective of diversity, representation, and inclusion. What do you notice about its character demographics and story choices? To what extent does it engage with diversity and how well does it succeed in doing so? How has the game worked to include other perspectives and viewpoints? What problems or concerns do you notice? Then, do a bit of online research and see how other player communities have responded to this game. How do players of

different demographics than your own feel about representation in this game? What concerns have others raised about representation?

(2) Create a short game using the platform of your choice to depict an experience or perspective different than your own. Spend time researching this experience, maybe even by looking for online player communities geared specifically toward players of this demographic. For instance, if you are cisgender and hoping to create a nuanced and complex gay character, you might look for online player communities dedicated to supporting LGBTQ+ games and players. What kinds of storytelling and representation choices do these communities appreciate most? What elements of this experience do you need to understand in order to convey this experience effectively? Spend some time, too, exploring other creative works that have dealt with this experience. What are some common pitfalls and problems? What approaches have worked well for other creators? Think too about game design tools and the kinds of experiences that are only possible within games. How might you use game mechanics or design choices to help others understand elements of this experience? Once you've created your game, spend some time playtesting it, making sure to find testers who are members of the community you are trying to represent. Listen carefully to the feedback you receive, and ask questions if you don't understand. What can you do to make your game even more successful at representing this perspective or experience?

11

Designing Games for Empathy and Change

The Wide and Varied World of Game Objectives

Many of the games we've discussed in the book focus on fictional story experiences with things like made-up characters, fantastical worlds, magical powers, or terrifying monsters. As writers, we may be drawn to games for their capacity for storytelling, for the interactive elements that make these narratives vibrant and compelling, more engaging even, than stories told in more traditional print or film formats. In this chapter, though, we'll be focusing on games that have goals beyond fictional storytelling. While narrative may be important to these games, they're also engaged with something more concrete and challenging: evoking and inspiring real-world change.

Games in this category differ widely in form, approach, and objectives. Some, like the game *Spent*, which was created pro bono by an advertising firm for the Urban Ministries of Durham, want to develop public awareness of real-world problems like debt, poverty, and homelessness while also raising money for charitable organizations. Others, like *The Fiscal Ship*, created by the Serious Games Initiative at the Woodrow Wilson Center for Scholars, hope that their game helps players learn more about the complex process of creating and maintaining a balanced government budget. Still others, like *Freshman Year* by Nina Freeman or *Dys4ia* by Anna Anthropy, hope that their games encourage players to understand and empathize with difficult personal experiences such as sexual assault or gender dysphoria. Other games may emphasize innovative gameplay as a way of fostering unique storytelling experiences, such as *The Vale: Shadow of the Crown*, which uses 3D audio and haptic features to create a gameplay experience accessible by the visually impaired. While these games may vary in content and approach, they all share a focus on real-world engagement and the belief that games can communicate ideas and experiences in unprecedented new ways.

An examination of these kinds of games can teach us skills that are useful in a variety of game writing circumstances. This chapter will explore helpful approaches for using games to communicate complex real-world issues, including strategies for developing and establishing empathy among your audience. We'll also discuss how you can use the game design elements discussed in previous chapters to help players understand personal experiences and to engage audiences, even those who might initially be resistant to the ideas you're exploring. How might we use game writing approaches for expressing a variety of real-world situations and experiences? This chapter will begin to answer that question and more.

Games as Practical and Rhetorical Tools

In the popular book *Reality Is Broken*, written by Jane McGonigal and published in 2010, McGonigal addresses a key concern that people often express when talking about video games—namely, that games are addictive and manipulative, and that the amount of time players spend with them has the potential to cause lasting social harm. This is a common argument and one that as game players we've probably heard expressed by a variety of people in many different contexts. At its heart lies the assumption that games are a dangerous and escapist waste of time, that they lead to a variety of problems such as increased propensity for violence, lack of ambition, or underdeveloped social skills, and that engagement with games should be tempered by a great deal of caution and guidance. For game players and writers, these arguments can be challenging to navigate. We enjoy games and the experiences they afford, but we certainly don't want to harm ourselves or the audiences who might play the games we create. It can also seem like the people making these arguments may be doing so without fully understanding games or even playing them. If they would only just listen, maybe they would be less afraid of some of the perceived dangers of games and technology.

In *Reality Is Broken*, McGonigal offers a different perspective on some of the so-called dangers of games. She encourages readers to consider an alternate possibility, mainly, that games offer challenging, satisfying experiences that are often missing from our day-to-day lives. Rather than seeing gaming experiences as time wasted, she argues that we should instead consider them an investment in improved skills and collaborative problem-solving, abilities and competencies that offer important social value. McGonigal writes, "The real world just doesn't offer up as easily the carefully designed pleasures, the thrilling challenges, and the powerful social bonding afforded by virtual environments. Reality doesn't motivate us as effectively. Reality isn't engineered to maximize our potential. Reality wasn't designed from the bottom up to make us happy." It's common to look at games primarily as commercial ventures, designed to maximize the time and money that players spend using them, exploiting audiences to increase the profits of large corporations, and McGonigal does not expect us to ignore the commercial and capitalist elements inherent in the game industry. Rather, she suggests that instead of rejecting the power that games offer, we instead study them, and consider how game design approaches and concepts might help us to make "reality" more rewarding and satisfying.

McGonigal encourages us to move beyond apocalyptic thinking and instead asks what games can teach us. If games are fulfilling genuine human needs that may be difficult to satisfy otherwise, how can we learn from games and use this model of engagement as a way of improving the world and our individual experiences? McGonigal's vision for the future of games is a bit different than what we might be accustomed to hearing. She writes:

I foresee games that make us wake up in the morning and feel thrilled to start our day. I foresee games that reduce our stress at work and dramatically increase our career satisfaction. I foresee games that fix our educational systems. I foresee games that treat depression, obesity, anxiety, and attention deficit disorder. I foresee games that help the elderly feel engaged and socially connected. I foresee games that raise rates of democratic participation. I foresee games that tackle global-scale problems like climate change and poverty. In short, I foresee games that augment our most essential human capabilities—to be happy, resilient, creative—and empower us to change the world in meaningful ways.

The majority of McGonigal's book examines games that have been designed with the aim of creating a real-world impact, such as *Foldit*, a free protein folding computer game developed by university scientists, which has helped researchers make discoveries about a variety of contemporary biological issues, such as influenza and Covid-19, and has also been helpful in drug development and other medical findings. McGonigal also talks about the process and approach for creating her own game *Superbetter*, a real-world game designed to help her cope with a difficult concussion injury and one that is meant to help players set and meet important goals, improve overall happiness, and foster strong mental health. For McGonigal, games have the potential to support lasting and important changes, which can help to make the world a better place.

Like McGonigal, James Paul Gee, an American linguist with a background in literacy studies, argues that integrating game design strategies into educational approaches can help improve student learning. Beyond arguing for the use of games in the classroom, Gee encourages educators to consider how game design approaches might help them better meet student learning goals. Gee argues that games are primarily tools of learning, as players must grapple with complex problems of increasing difficulty. Because most games are for-profit ventures, Gee maintains that learning in games must be fun and compelling, encouraging players to replay in the place of failure, to build detailed and complicated banks of specific knowledge, and to collaborate with others while doing so. He points out that most game players are self-directed and motivated, spending hours of time memorizing intricate details and consulting (and creating) online resources such as wikis and forums. In his 2003 book *What Video Games Have to Teach Us About Learning and Literacy*, Gee argues that integrating game design principles into educational systems could revolutionize approaches to teaching, creating a strong and lasting impact on student learning.

Beyond arguing that games and game design concepts offer the potential to affect real-world change, Ian Bogost, in his 2007 book *Persuasive Games: The Expressive Power of Videogames*, maintains that games are powerful rhetorical tools with the potential to persuade and compel audiences in ways that are different and more formidable than other media. As a scholar, Bogost is interested in the ways that games have the potential to act as complex systems. As an example, he examines the *McDonald's Videogame*, released in 2006 by Molleindustria, a pseudonym for Paolo Pedercini, who creates games that focus on contemporary cultural issues. In the *McDonald's Game* (unsurprisingly not sanctioned by McDonald's), players must plant soy, raise and slaughter cattle, hire restaurant workers, and make decisions on larger corporate policies, thus creating a working small-scale model of the fast-food industry. Molleindustria's game offers players options for dubious ethical choices, such as advertising to children, bulking up the animal feed with questionable ingredients, or bribing public officials, all of which make it easier to excel at the game. Thus, the game not only models the systemic complexity of the industry, but it also presents a powerful critique of problematic industry practices, all in a twenty-minute experience that is accessible to players of all ages. Bogost's analysis of the *McDonald's Videogame* serves as an example of how video games offer the capacity to make complex real-world arguments in unprecedented ways, providing rhetorical approaches that are understudied and often disregarded or ignored.

Further, Bogost argues that all games are ideological tools, whether this is done intentionally or unintentionally by their creators. Because games model and act as systems, and because those systems are created by designers, all games require players to participate in and adopt interconnected rules, which are shaped by preconceptions. Bogost (2008) writes, "No video game is produced in a cultural vacuum. All bear the biases of their creators. Video games can help shed light on these ideological biases. Sometimes these biases are inadvertent and deeply hidden. Other times, the artifacts themselves hope to expose their creators' biases as positive ones, but which of course can then be read in support of opposition." In Bogost's view, we should all be aware of the ways that games act as powerful persuasive tools that encourage (and often require) players to act in certain ways or to adopt particular perspectives, and that the rhetorical strategies of these tools may be especially compelling and convincing.

Bogost's book presents a direct response to the argument that video games are an escapist waste of time and unworthy of attention, but it offers a more sobering perspective than McGonigal's idealistic claim that embracing game design principles will automatically improve the world. Dismissing games as a medium is dangerous, because, as Bogost writes,

> We always play when we use video games, but the sort of play that we perform is not always the stuff of leisure. Rather, when we play, we explore the possibility space of a set of rules—we learn to understand and evaluate a game's meaning. Video games make arguments about how social or cultural systems work in the world—or how they could work, or don't work. . . . When we play video games, we can interpret these arguments and consider their place in our lives. (2008)

When left unexamined, ignored, or misunderstood, the rhetorical strategies of video games are particularly powerful and have, like advertising approaches, a great potential to shape and alter human behavior. Rather than dismissing games, Bogost encourages parents and educators (and everyone!) to instead learn more about them by looking at them critically, particularly with the goal of teaching children and young people to understand these unique persuasive strategies, as both consumers and creators.

As game writers, Bogost's ideas are particularly important, as they encourage us to consider our own ideologies and perspectives, and the ideas and worldviews that we may inevitably write into the games we create, often without even realizing it. How are our own implicit biases and preconceptions affecting or shaping our creative projects? What rhetorical strategies can video games offer that might help us to make our ideas even more compelling and persuasive? What does critical examination reveal about the ideological underpinnings of our favorite games and does it push us to see them differently? Bogost writes:

> Video games are not mere trifles, artifacts created only to distract or amuse. But they are also not automatically rich, sophisticated statements about the world around us. Video games have the power to make arguments, to persuade, to express ideas. But they do not do so inevitably. As we evolve our relationship with video games, one of the most important steps we can take is to learn to play them critically, to suss out the meaning they carry, both on and under the surface. (2008)

Beyond considering the games we play from a more critical perspective, Bogost's ideas push us to look closely at ourselves and the pieces we create. What are we choosing to create, and what kind of impact would we like for it to have? Are we aware of the perspectives and views we may be inadvertently reproducing in our own works? To what extent do we want to engage with real-world change in our games, and how can we use the unique rhetorical strategies of games to do so? We may, as McGonigal suggests, create games with the goal of improving the world and fostering change, but Bogost proposes that our games may have a great power to persuade and shape human behavior, even if we set out with less lofty goals.

Case Study: *Night in the Woods* (2017)

In the 1970s and 1980s, working-class fiction exploded in popularity. Wonderful writers like Raymond Carver, Andre Dubus, Sandra Cisneros, James Alan McPherson, Bobbie Ann Mason, and Breece D'J Pancake took a microscopic lens to the lives of beer-and-shot blue-collar workers who are rarely protagonists in high art. Class consciousness and even rote depictions of working-class life are exceedingly rare in video games and that's part of what makes *Night in the Woods* so meaningful to so many.

Night in the Woods is set in a fictional rust belt town that very much resembles the kind of down-and-out rural areas scattered around Pittsburgh where two of the game's three designers live. You play as Mae Borowski, a twenty-year-old college dropout who returns to her postindustrial hometown. Mundane moments spent reconnecting with friends and family comprise the bulk of the game. Parents are saddled with predatory mortgages and fear losing their homes. The town's youth struggle under service jobs that leave them exhausted and depressed. The malls, mines, and churches are abandoned, but the bar's always packed, filled with depressed locals cheering their football heroes. A father transitions from working full-time in a factory to hourly in a supermarket. Angry neighbors write poems about the brutality of capitalism and burning Silicon Valley to the ground.

The power of *Night in the Woods* lies in how it grounds its themes not in the worn-down characters of Carver but in the queer millennials of Mae's generation. The game replicates the decaying urban landscapes of Andre Dubus while borrowing tonally from offbeat graphics novels like the *Scott Pilgrim vs. the World* series or even the surrealism of novelist Haruki Murakami. Accentuating this stylized dramatization of the evils of capitalism is that every character is rendered as a cartoony animal. This aesthetic allows the game to toggle rapidly between absurdity and despair. In one moment, you

Figure 12 It's fallen to Mae Borowski to save her family from economic ruin.

help a queer anarchist break into a children's museum while shouting "Crimes!" In the next, you witness Bea—the game's emotional anchor—drive ninety minutes to a college party where she can pretend to be middle class with a real chance at upward mobility. It's a heartbreaking scene that's amplified when Mae reveals to everyone that Bea's not a college student and actually runs a hardware store half a state away (Figure 12).

In a media landscape where honest depictions of working-class people are rare at best, *Night in the Woods* is a singular experience that illuminates what it means to be a working-class millennial trapped in the service industry instead of the mines and factories of their parents. Its dialogue and characterization stand shoulder-to-shoulder with the finest literary fiction, and it's a key example for anyone looking to write about class at any level in any medium—but especially in a video game.

Diary-Style Games and Personal Narratives

In the game *Vivant Ludi* by designer Caelyn Sandel, players are asked to step into the shoes of Vivian Grimes, an indie game developer who creates games from home. Vivian has a project due in five days, but she is struggling to work while maintaining a regular self-care regime and dealing with harassment. In this short game, which takes place entirely in the rooms of Vivian's small apartment, players must help Vivian do basic daily activities like eating, showering, and going to bed on time, while also helping her to work on her game and take occasional breaks by talking to her online friends. The game is visual, and players move between each room, clicking on things like Vivian's bills and her computer. Many tasks are repetitive, such as eating and showering, and they result in the same text each time the player selects them. At the end, Vivian releases her game, which is met with undistinguished response. On the surface, *Vivant Ludi* may not fit with our assumptions of what a game should be and how it should work. The actions within the game are uneventful: eating, sleeping, showering, feeding the goldfish. Very little changes during the course of the game, which is impossible to fail or lose. If Vivian doesn't shower or sleep, she simply feels bad. Even if you neglect to feed the goldfish, it will never die. The end of the game feels uneventful, and it doesn't include the rising tension that we often expect from stories and that is described earlier in this book. It's not uncommon to finish a playthrough of *Vivant Ludi* and think, "Is this all?"

When examined as a diary game, however, it's clear that *Vivant Ludi* has entirely different goals from what we might associate with a typical game. In a talk called "Rise of the Diary Game," given in 2014 at Alter Conf, a conference dedicated to highlighting marginalized voices in technology and games, Sandel describes the idea of the diary game and explains how it relates to her own work as a game designer. Unlike more mainstream games, diary games tend to be smaller, often created by one person or a small team of people, and their main goal is to portray a personal experience or a fictionalized version of a personal experience. In Sandel's view, the purpose of this kind of game is to convey emotion and encourage empathy in a way that is difficult to do in other media, and often these kinds of games express experiences or viewpoints that are less common in more mainstream media, educating players about personal experiences they might not encounter otherwise. While this kind of game often succeeds in reaching and inspiring others, particularly those who are unaccustomed to seeing their own lifestyles and viewpoints in games, creating these games is often an act of catharsis for the author and sometimes working personally through a difficult experience is the central goal of a game's creator.

Sandel created *Vivant Ludi* as an entry for Ruin Jam, a game jam created in 2014 in response to Gamergate and the backlash against small indie games that highlight diverse perspectives. Ruin Jam was dedicated to amplifying marginalized voices, advertising itself as "open to anyone and everyone who has been, is being, or plans to be accused of ruining the games industry." In her Alter Conf talk, Sandel makes a case for games that have different goals, arguing that diary games may lack some of the things associated with more mainstream games. They may feel less fun to play, or deal with difficult subjects, or not be as interactive, but this is often by design. The primary goal of a game like *Vivant Ludi* is not to entertain players as much as it is to convey a deeply personal experience. In helping Vivian to manage her self-care and complete her creative work, the game asks players to participate in a lifestyle that may be very different than their own, engaging with mundane tasks and everyday challenges in a way that requires them not only to empathize with this experience but also to actually enact and participate in it. Sandel argues that many people turn to games for an escape, and diary games may be particularly disruptive because they lack that sense of comfort and fantasy. Instead, the game mechanics may make players feel trapped, helpless, or bored, as a way of attempting to communicate unpleasantness, or even implicate players in systems of oppression.

For Sandel, the capacity for diary games to disrupt the status quo is precisely the reason they came under such harsh scrutiny during Gamergate. Unlike other forms of media, games require participation, engagement, and empathy, and often diary games may ask players to identify with perspectives they find challenging or uncomfortable, or may draw awareness to the privileged roles we play in problematic systems, experiences that can be deeply unpleasant. Sandel says, "A privileged gamer doesn't want to be aware of their privilege, or their complicity in oppressive structures, especially in the environment [games] they consider neutral. They don't consider their neutrality political. They don't want to be told that their neutrality is political." Sandel, like Bogost, notes the potential for games to communicate alternate perspectives and ideologies in a way that may be deeper and more compelling than more traditional media experiences. Ultimately, though, for Sandel, diary games like *Vivant Ludi* are less about player experience and more about the developers making them. These kinds of games offer "an opportunity to boost the voices of the marginalized in their own words and images, without speaking over them." In this way, diary games offer the potential for hope and change in the world by acting as a space that empowers and encourages other voices, particularly those that might be absent from more mainstream media.

Transmedia, Alternate Reality Games, and Audience Engagement

In June 2006, sixteen-year-old Bree Avery, homeschooled and bored, began posting videos to YouTube under the username lonelygirl15 as a way of reaching out and finding an online community. Initially, Bree's videos were short, silly perspectives into her insulated life, focusing on her experiences studying, her purple monkey puppet, and Daniel, her closest friend. Over time, though, Bree became more open about her family's religion and increasingly worried about its connections to a dangerous cult. She and Daniel reached out to viewers through YouTube and other websites like MySpace (a social media site popular at the time), asking for help and responding to audience suggestions and comments. When viewers suggested that Daniel might have a crush on Bree, she responded, surprised, in a subsequent video. As she struggled to rebel against the potentially dangerous rituals of her family's religion, viewers were there to support her, offering advice and encouragement. Viewers helped the pair solve puzzles and riddles, sharing insights with one another on forums they created and ran themselves, discussing episodes and pouring over details. By September 2006, lonelygirl15 achieved the record for being YouTube's most subscribed channel, a distinction it held for over 200 days. By this time, most fans had learned what you've probably already guessed: lonelygirl15 is not an actual teen video blog but a scripted series starring actress Jessica Rose and one of the first, and most popular, alternate reality games ever created.

Alternate reality games can take many forms, but most use the internet to tell an interactive story that takes place in real time, using videos, websites, social media, and other online communication approaches to adapt to viewer feedback and response, creating a large-scale narrative that relies on audience participation. In his 2006 book *Convergence Culture*, Henry Jenkins explores the idea of transmedia storytelling, a perspective on narrative that is arguably central to some of our most powerful and popular contemporary stories. The term "transmedia" applies to any story that is told across platforms and genres, allowing the narrative to expand and develop in a variety of contexts, and providing audiences access to it from multiple entry points.

If you think of any massively popular narrative from the last few decades, it is likely a transmedia story. The stories of the *Star Wars* franchise, for instance, take place in movies, television series, novels, comics, video games, role-playing games, and other venues. A *Star Wars* fan might first enter the world by watching *The Empire Strikes Back*, or by streaming *The Clone Wars*, or by reading the *Rogue One* graphic novel. Each of these works stands on its own, but exploring all of them together creates a detailed, complex world full of stories and perspectives that is larger than any one narrative, and too much for any one person to hope to ever consume or remember. Jenkins calls this "collective consciousness," arguing that the internet allows us to work together to curate colossal banks of knowledge and information that in turn enables storytellers to create massive, multifaceted worlds that are larger in scope than something any one person could ever hope to design or understand. In short, technology and the internet enable us to create immense stories, and many of our most popular contemporary narratives (*The Matrix*, *Spider-Man*, *Harry Potter*, *Game of Thrones*) fall into the category of transmedia.

While the term "transmedia" can apply to multiple stories that stand on their own and take place in multiple platforms and genres, transmedia can also apply to a narrative like lonelygirl15, meaning one story told across a variety of online platforms. These kinds of transmedia stories often expect viewers to visit websites, watch videos, and view social media profiles, and they often adapt to and incorporate audience feedback into the narrative. In the case of this kind of narrative, each individual piece (a character's Twitter feed, for example) might not stand on its own as a complete story, but it still acts as an entry point for the narrative and creates a sense that the story is happening in real time, right now, and that the audience is a part of it. In the transmedia web series *The New Adventures of Peter and Wendy*, for instance— adapted from J. M. Barrie's novel *Peter and Wendy* and released in 2014— YouTube videos follow contemporary characters based on classic characters from *Peter Pan* as they work at a newspaper, the *Kensington Chronicle*. The characters have social media profiles and Twitter feeds, and the newspaper has its own website, where viewers can post profiles to a dating service, view classified ads, or write questions for Wendy, who works as an advice columnist. Unlike lonelygirl15, this series is clearly fictional, but it still relies on audience participation and engagement, often integrating viewer feedback into the narrative. The characters respond to viewer questions and comments, and the story is broad and complex, taking place on a variety of platforms, all of which contribute something unique to the narrative.

Transmedia narratives and the alternate reality games that often rely on them are interesting and complex and require more space than we have time to dedicate here. For game writers, though, these approaches are useful and worth considering. For one thing, transmedia narratives rely on audience engagement and feedback, often adjusting plot points or approaches based on viewer response. They may ask participants to engage with the story directly by solving puzzles or deciphering clues, and they can even ask audiences to engage in real-world action, traveling to a specific geographical place, for instance, or attending an event. As a result, these kinds of game narratives have the potential to blur the boundaries between the fictional world and the real one, sometimes, as in the case of lonelygirl15, pushing viewers to confuse fact and fiction. Many games rely on complex social mechanisms, and as a result, players often form close communities, which frequently surpass the confines of the game. In a transmedia narrative, these connections are an integral part of the story. Without viewers to comment and respond, the story could not progress and so compelling audience engagement is a crucial part of the writing process. Similarly, writers must be willing to adapt their narratives, sometimes at a moment's notice, and to relinquish some degree of control, as it can be difficult to predict where participants will enter the narrative and how they will engage with it. The reward for dealing with these challenges as a writer is a story that is closely enmeshed with real-life experiences, one that may foster lasting long-term relationships between its participants and that has the capacity to inspire a strong sense of personal connection with the game and the story. Transmedia stories and alternate reality games can be powerful, and practicing these techniques will help you develop useful strengths as a writer.

Tips and Techniques

Approaches to writing games that engage closely with the real world are varied and will depend a great deal on the project you have planned. There are a few skills, though, that will help you in a variety of situations. One of the first techniques to practice is to consider and incorporate Ian Bogost's ideas of ideology into the games you are making. What kinds of messages, ideas, or situations would you like players to encounter? What kinds of actions will bring players to a closer understanding of a particular system or situation, perhaps even implicating them in it? To what extent would you like for audiences to question their own roles in the subject you are describing? If you are writing a personal game, like *Vivant Ludi*, you'll want to consider the actions that are most central to the character's experience, in this case, eating, showering, working, and sleeping. If you're critiquing a system, like in the *McDonald's Game*, it's useful to create questionable choices that have lucrative rewards and lead to even more dubious choices and situations. Forcing players to hurry can often push them toward making unethical choices, as in *Papers, Please*, where players must process a certain number of passports and identification materials within a time limit each day in order to afford rent and feed their families. In the beginning, it may feel easy enough to ask personal questions in the press of clamoring travelers, but what about when you're offered a bribe or asked to use a camera that can see through a person's clothing? Where will you draw the line? Will you draw a line at all? Carefully planning the sequence of choices and actions, crafting scenarios with consequences that are difficult to predict or avoid, and doing your best to be conscious of the ideological underpinnings of the story you are telling are useful ways to enmesh players into difficult and uncomfortable scenarios.

In games focused on real-world issues, developed research and planning is of crucial importance. Obviously, if you're creating a game based on folding proteins like *Foldit* or helping people to learn about government finances like *The Fiscal Ship*, you want to make sure that the material you're including is as accurate as possible. Beyond this, though, there are a number of things to consider when writing this kind of serious game. For one thing, you'll want to consider your goals and imagined audience. Are you hoping to teach people? Encourage them to donate money or change behavior? Foster empathy with a particular experience or viewpoint? Your approach to game writing may change depending on your primary goals. It's also important to consider your audience. Whom do you imagine will be playing your game? What kinds of information will they need to understand? What barriers might prevent them from engaging with the topic, and how might you address those barriers? Depending on the audience you have in mind, you may need to tailor the amount of research you include so as not to overwhelm them. The game *Spent*, for instance, meant to help players understand more about poverty and homelessness, could easily provide players with extensive statistics and case studies. Instead, the developers chose to incorporate only a few carefully chosen facts, interspersing them throughout the narrative to have the most impact.

Depending on your subject matter, you may choose to incorporate fictional characters or story lines to help engage viewers and encourage them to connect to the material. In this case, you'll need to plan carefully to make sure that your story and characters are engaging, but that they augment the information you want to convey rather than distracting from it. Similarly, you'll want to think about the emotional impact of the subject you're dealing with and consider the ways that story will work with that tone. In the 2016 game *Liyla and the Shadows of War*, for instance, creator Rasheed Abueideh hopes to give audiences insight into the Palestinian perspective on the conflicts in the Middle East. The game's focus on Liyla and her family's struggle to escape an attack on their neighborhood uses a fictional story in conjunction with carefully chosen facts to engage players' emotions and help them empathize with perspectives that are often absent from mainstream media.

Fostering empathy and connection between players and characters is a challenging skill and one that is particularly important for diary games, which often focus on personal experience. Players can connect to a character through cognitive empathy—by understanding another person's mental state or perspective—or through emotional empathy—by responding to another person's emotions. Often this works best if players are asked to connect with characters right from the beginning and if they have a specific idea of how actions and choices influence issues in the game through clear feedback. Empathy can be easier to develop in short bursts, particularly if connecting with a character might challenge a player's belief systems—*Liyla and the Shadows of War*, for instance, only takes about fifteen minutes to play. It's also useful to emphasize points of similarity between the player and the character, and to provide quieter moments, which give players the mental space to process the story and connect with it. The central characters in *Liyla and the Shadows of War* are a father and daughter, a family relationship that will resonate with most players. Though much of the game is spent avoiding gunfire and running from missiles, the game also includes breaks in the action that allow players to focus on the relationship between Liyla and her family. Fostering an emotional response, such as empathy, might seem difficult to engineer, but if this kind of connection is important to your game, it's useful to consider strategies and approaches that can make that experience more likely.

Finally, examining alternate reality games or transmedia stories can help illustrate useful strategies for engaging with audiences. In this kind of game, it's important to establish a clear timeline. When does the story start and how long will it last? If your story will take place over multiple platforms, you'll need clear mechanisms for directing audiences to the other pieces, which often takes place within the narrative itself. Perhaps a central character will post a video making note of something that happened on a social media site, like Twitter or Facebook, or perhaps one character will link to another character's profile page or video blog. Recognize, too, that different platforms convey different kinds of experiences. Social media pages are useful to help your audience learn more about characters and connect to them, while more formal platforms, like websites, are great for establishing information about the larger elements of the world. It's useful to consider the beginning of your story carefully. What kind of inciting action or event will draw in participants? You'll also need to prepare a fair amount of backstory material before the inciting incident to give the sense that the characters are real people with full lives and relationships.

Part of the fun of creating an alternate reality game is making the audience a part of your story. You'll need to teach your audience how to engage with your creation by training them where to look for important story materials. If you plan to incorporate audience feedback into the narrative, make sure to set a precedent by letting the audience know that their comments and posts are being read and responded to by the characters. You may want to start with incorporating smaller moments of audience input into the story before moving to larger ones. Similarly, if you'd like your audiences to help with solving a puzzle, you may need to help them recognize what is expected of them and to provide clues if they struggle to locate important information or to know what to do with it. Successful transmedia storytelling often relies on careful planning and extensive preparation, but you'll also want to incorporate a fair amount of flexibility to respond to your audience and allow them to shape the narrative.

Overview

The world of game narratives is diverse and vast, and though games are frequently charged with causing social problems, it is just as possible that the games you create might have a positive, lasting real-world impact. Games have the potential to engage audiences, encourage learning, establish communities, and promote empathy, but these effects often won't happen without careful planning and attention to design choices. Whether you are working to educate your audiences about a complex social issue, to help them understand an often-ignored perspective, or to draw them into a fictional story that might easily be confused for reality, games have the capacity to support learning and bring about change in a way that is difficult in any other medium. The more you are able to practice and refine these strategies, the more prepared you will be for using your game writing skills to change the world.

Exercises

(1) Choose a real-world issue that you feel is important, perhaps one that is often neglected in mainstream media. If you need to learn more

about this issue, take some time to research it before continuing this exercise. Then, plan a game that will explore this issue by doing the following:

(a) Imagine your audience. What would you like for them to know or understand about this topic? What do you expect they will know already? What barriers or misconceptions might make it difficult to learn about this subject?

(b) Consider your rhetorical approach. Will you base your game on real-world scenarios or focus on a fictional story? To what extent will you incorporate research and where will you include it?

(c) Make a plan for interactivity. What kinds of choices would you most like for players to be able to make in this game? For instance, a game about sustainability might ask players to make choices about purchasing things as well as about throwing them away. And in the game *Spent*, players must make choices for themselves, but they must also consider choices on behalf of their child.

(d) Consider all possible outcomes and endings. What kinds of information would you like players to learn? What kinds of endings should be possible? To what extent will you incorporate failure and replay and what will that look like?

(e) Make your game! Choose whatever platform you prefer. (If you're not sure which platform to use, you may want to look over Chapter 12 on tool suites.) Then playtest it. How accurately have you judged your audience? Have they learned the information you were hoping they would learn?

(2) Choose an everyday experience, one that you know well. Then, create a short game using the platform of your choice to approximate it for people who might not be familiar with it. The issue you select can be serious or humorous, but whatever you choose, it should use game design techniques to convey reality. (Students who have completed this exercise in the past have focused on things like housetraining a puppy, teaching children to swim, dealing with anxiety or another mental health issue, changing an insulin pump, auditioning for a school play, or training for a race.) Make a list of the kinds of choices individuals experiencing this often deal with. What kinds of choices would you like players to make? Consider failure and replay, too.

Should there be multiple endings, or should all paths lead to the same place? Then, make your game and playtest it! Have your players understood this experience? What might you do to help make this experience even more clear to a potential audience?

Part III

Practical Tools and Approaches

12

Tool Suites and the Game Industry

From Words to Games

By this point, you've learned a lot about creative writing, game design, and the obstacles and opportunities associated with writing a video game as opposed to static media. Perhaps you've already completed some of the previous exercises in this book and have a few Word documents on your computer filled with ideas and dialogue and choices. However, no one can play a Word document, and game companies aren't always willing to evaluate scripts when hiring writers. What if we want to self-publish a game that speaks to power or to our own communities? What if we want to curate a portfolio of playable work to include with job applications?

This chapter will cover that, which means learning the basics of four specific tool suites, not to mention what to do with your finished games when you're done. This instruction is intended for the student/teacher, hobbyist, and writer looking to break into the games industry as a professional. However, video game tool suites are always changing and adapting to the needs of players and creators. Going too in-depth in these sections would almost immediately make this chapter out-of-date, and, instead, we'll provide a broader survey intended to get folks started. This will include the assignments we use in our classes, but we encourage you to search for further resources online if you're interested in using the more advanced functions of any of these free programs.

Twine

Released in 2009 by Chris Klimas, a Baltimore-based web developer, Twine is a wonderfully robust tool that allows writers to generate choose-your-own-adventure games that are playable in almost any web browser. Twine is one of the easiest programs for students and writers to learn, and within minutes almost anyone can write a story that includes a few choices and is playable in a webpage. However, for the more technically advanced user, there's a lot under the hood to tinker with. Although completely unnecessary if you want to build a simple game, users can add Javascript, HTML, or CSS to their Twine games. This allows designers to incorporate everything from music, images, countdown clocks, health meters, and more. Plenty of users post their code to Twine forums when they're done, so it's easy to copy/paste these tools into your own games even if you have no experience with coding yourself.

One of the most famous Twine games is 2013's *Depression Quest* written by Zoë Quinn, which we've discussed in detail throughout this book. Two other examples we often show students to demonstrate Twine's capabilities are 2015's *The Writer Will Do Something* and 2016's *Queers in Love at the End of the World*. The former, cowritten by Tom Bissell and Matthew S. Burns, sees the player assuming control of a writer of an AAA video game who must navigate a contentious story meeting. Obviously, this is a meta premise straight out of the aforementioned *Stanley Parable* playbook, but *The Writer Will Do Something* plays it straight. Instead of providing gonzo choices and outrageous outcomes, *TWWDS* relies on an encyclopedia structure. There are choices, but more frequently a character's name will be highlighted instead of some semi-related backstory or event. Clicking these links rarely sends the player down a branching path. Instead, these links reveal additional backstory or context. Students immediately imagine reading a traditional novel like Zadie Smith's *White Teeth* and clicking the name of a background server in the Palace, revealing a litany of unrevealed context and characterization. In contrast, *Queers in Love at the End of the World* written by Anna Anthropy is a rapid, micro experience. The world is ending in ten seconds and the player must decide whether to kiss her girlfriend, take her hand, hold her, or tell her. What that "tell" is won't be revealed until the player clicks it, but an ever-present clock on the left of the screen ticks away in real time to zero at which point "Everything is wiped away." Here, students see the possibilities of including a tiny bit of code—the countdown clock—to tell a story that would be impossible to render in static literature.

After our students have played a few Twine games and discussed them, we begin showing them the software in a large group demonstration. As of this writing—and for more than a decade—writers can access Twine and copious reference information at https://twinery.org/. Twine can be run from either inside a writer's browser, or you can download it on Windows, Mac, or Linux computers. When you first boot up either version of Twine, you select the option on the right-hand side to create a brand-new story, name it, and then you're brought to the main workspace. Here, you can access a number of story cells, each one representing a page of text the player can read. We often ask our students to experiment here, creating a simple paragraph of text with two simple choices. Here's a typical example:

> A hungry dog enters a room. She sees her kibble overflowing in her food bowl, but she also smells a delicious cherry pie cooling on the windowsill in the breeze. The dog saunters over, licking her chops.
>
> Bend low and eat the kibble.
>
> Pop up on your hind legs and steal the pie.

After creating a simple—and ridiculous—situation like this, we ask our students to test their games by pressing the "Test From Here" option on the toolbar. Almost instantaneously, Twine will produce your story into a simple HTML game, but the students will quickly recognize the problem—Twine doesn't know that the bottom two lines of the hungry dog story are supposed to be choices leading toward two branching paths.

This is when we often reassure the humanities students in the room—or any student in the room with no experience in coding. They will have to code to communicate to Twine that this kibble/pie dichotomy is supposed to be a choice, but this is the only code they will ever have to learn to make functional Twine games. We've used Twine in our classrooms for eight years—in a variety of settings with both traditional undergraduates and adult learners—and no one has ever failed to learn this simple code. Let's return to our hungry dog story and modify it slightly.

> A hungry dog enters a room. She sees her kibble overflowing in her food bowl, but she also smells a delicious cherry pie cooling on the windowsill in the breeze. The dog saunters over, licking her chops.
>
> [[Bend low and eat the kibble.|Kibble Path 1]]
>
> [[Pop up on your hind legs and steal the pie.|Pie Path 1]]

This is all it takes to tell Twine that you intend for these sentences to be choices that lead to branching paths—two brackets on the left, a pipe after the text, and two brackets on the right. It's important to note here that the text that comes after the pipe—"Kibble Path 1" and "Pie Path 1"—is never seen by the players. This text is purely for the writer, and Twine will immediately generate two new story cells called Kibble Path 1 and Pie Path 1 connected to our original cell by a line. Capitalization and punctuation need to match exactly, but otherwise, learning this simple code couldn't be easier (Figure 13). We repeatedly use this example to help students remember: [[what the player sees|where the player goes]].

Although advanced coders can employ some of the more complicated techniques we referenced earlier, you now know everything you need to write a fully playable game. Twine features a simple Publish to File command under the Build option in the toolbar—and we'll cover how to publicly release your game and how to build a portfolio later in this chapter. We'd suggest Twine to writers who want to focus on either a branching narrative with lots of choices or an encyclopedia game where the player can delve into as much backstory and context as they want. It's not well-suited for someone who wants to tell a linear story, and it's much harder to incorporate interactive graphics or music—although this can be done with enough coding expertise. Twine is the perfect tool for writers looking to dip their toes into writing games, and it's always the first tool we teach students in our game writing classes. Below, you'll find our first low-stakes Twine assignment and then you'll find our more formalized high-stakes assignment. The first ensures that students can successfully create a game with choices. The second asks them to write a game that strives for emotional resonance.

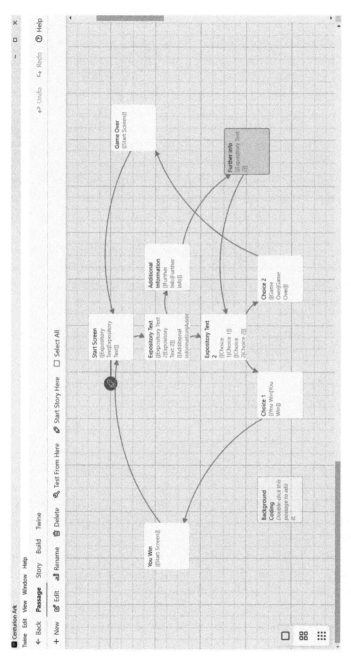

Figure 13 A typical Twine project layout.

Low-Stakes Twine Assignment

Pretend that you've just graduated. Rockstar Games, developer of *Grand Theft Auto* and *Red Dead Redemption*, puts out a call for a junior writer and you send in your resume. Great news! You've made the final round, and Rockstar flies you out to San Diego for the interview. **While there, you're asked to complete a writing test. The interviewer sits you in front of a computer with Twine installed**. They tell you to hypothetically imagine a gunslinger in the Old West. Said gunslinger has entered a dusty frontier street with mysterious intent. The gunslinger discovers an old saloon packed with some of the wildest and roughest hooligans imaginable. **The interviewer tells you to write three cells in Twine. Each cell should include two types of writing—description of what the player sees in the ensuing cutscene and also dialogue. In cell one, the player should be presented with a meaningful choice.** It's your job to write satisfying outcomes for both of these choices. You have twenty minutes. **Your future employment depends on your ability to generate a meaningful choice in a Western setting with two compelling outcomes for players.** You put your hands over the keyboard and begin.

To submit your project, you must hit Publish to File under Build, then save to your computer. Do not just copy/paste the html link. No one can use that but you.

High-Stakes Twine Assignment

Level 1: Assignment

When we think of narrative in video games, we often think of branching narratives. Although radically different in scope and tone, *Queers in Love at the End of the World*, *The Writer Will Do Something*, and *Depression Quest* all feature branching narratives where players must make choices that lead down slightly different or extremely different roads. Now that you're in the process of learning Twine, it's time to practice this ourselves. For this assignment, **you will produce a Twine game with a minimum of 2,500 words, at least four branching paths, and at least four different endings**. Thirty cells is a good number to aim for. Do not copy/paste cells. **We want to see new and fresh writing for every cell and choice, and we want to see you provide the player with meaningful choices**.

But what to write about? Later in the course, you will have the opportunity to cast off restraints and write a game focused on whatever you want. But for now, you will find below a table of scenarios, genres, and themes. **For this assignment, you must mix and match and select one scenario, one genre, and one theme from the table**. Remember, 2,500 words and four paths don't give you much narrative space to maneuver in. At this early stage of the course, try to focus on a smaller experience closer to *Queers in Love at the End of the World* or *Depression Quest* than something extremely large like *Skyrim*.

Level 2: Progress Reports

Each student will be expected to work on their Twine game during three class sessions. **At the end of each of these three classes, all students will be expected to post a 300-word Progress Report online, detailing exactly what you accomplished during class and all you still have to do**.

Level 3: Completion

When you've completed and tested your Twine game, **remember to export your game by using the Publish to File command in Twine. You will then create a personal page on itch.io—which we'll discuss at length later in this book—where you will upload and release your game**. Your final game at the end of the semester will also be hosted here. **Post the url of your Itch page with your game to our discussion board**. If you don't want to release your game to an audience outside this class, remember to password protect your game and to also provide that password for us.

Level 4: Pitch Session

After releasing your game, you will be required to deliver a two-minute pitch about your project to the class. Pretend that we are potential players. What is your game, and why should we play it? Don't go over two minutes.

Level 5: Rapid Peer Review

After everyone has submitted their game, we will assign each student another student's project. You must then write a 500-word Peer Review that will help them revise their game. First, very briefly **describe** what a student's work is about. This should be the shortest section. Second, tell us briefly about the game's **strengths**. Finally, finish with **prescription**, a paragraph explaining very specific elements of the game that still need work and ways in which your peer can revise their game for the better. This should be the longest section.

Level 6: Grading Criteria

A strong Twine game will engage critically with its chosen scenario, genre, and theme—aspects that should become evident to the player without having to be told. There will be meaningful branches and meaningful choices. Your prose will be well-organized without redundancies and errors. The game will be well-tested and bug-free. Your Itch page will be professional. All students may revise their game at the end of the semester, and all initial grades are provisional. You will receive detailed feedback on your game from your professor.

Points: (100 total)
50 = Twine Game
10 = Progress Report #1
10 = Progress Report #2
10 = Progress Report #3
10 = Pitch
10 = Peer Review

Scenario	Genre	Theme
A thief breaks into a building.	Romance	Sustainability
Someone tries to get over a break-up in a single day.	Adventure	Faith
A character climbs a mountain and discovers something surprising.	Crime	Depression or Anxiety
A character returns home and discovers something surprising.	Sci-Fi	Gender
A character hears a strange noise in their basement.	Fantasy	Family
A character can time travel to the past but can only stay there for a single day.	Humor	Revenge
A character is exiled from their home.	Western	Mythology
A routine day at work goes very wrong.	War/Espionage	Hope
A vacation goes very wrong.	Horror	Compassion
All the clocks in a character's apartment have stopped.	Superheroes	Mercy
A character is imprisoned and tries to escape.		Apocalypse
A character's best friend has gone missing.		Friendship
A character has one day to retrieve a stolen item.		Childhood
It's a character's first day at a new job.		Cooperation
This first date goes very, very poorly.		Greed

It's vitally important to note here that we always include the milestones we've laid out above—Levels 2 through 6 specifically—in *all* of the high-stakes video game writing assignments we've included in this chapter. However, we haven't repeated them below in the sample assignments. For the remaining assignments, we'll only include instructions that are unique to the project.

A few further resources on Twine:

> https://twinery.org/cookbook/
> *Writing Interactive Fiction with Twine*, Melissa Ford (2016)

Audacity

The first thing you should know about Audacity is that unlike the other tools we're discussing, it's not strictly a video game development suite. Instead, Audacity is a audio-editing software that allows people to create everything from podcasts to songs. In our classrooms, we use Audacity as a fun way to teach students how to write audio logs. Earlier in this textbook we discussed walking simulators, games which rely on environmental storytelling where the player wanders around a space often listening to diary entries like in the seminal but retroactively problematic *Gone Home*. But examples of audio logs can be seen in many different kinds of video games. *BioShock*, a first-person shooter set in an Ayn Rand-inspired underwater hell, scatters audio logs across its dense map. *Everything* (2017) drops in real-world philosophy lectures from the writer Alan Watts. We use Audacity in our classrooms as a quick way to teach students how to produce their own audio logs, but it's important to note here that a game writer or narrative designer working in the industry wouldn't often, if ever, be expected to handle audio production themselves. This skillset is less about assembling a work portfolio to enter the games industry and more about creative expression, asking students to direct voice actors or perform audio logs themselves, talking to the kinds of communities they wish to address and speak power to.

To prepare students to write audio logs, we first ask them to listen to audio logs. Luckily, a huge amount of audio logs has been uploaded to YouTube by fans. On our learning management software class sites, we link to audio logs from a wide range of video game genres—the aforementioned *BioShock* and *Everything*, but also *Marvel's Spider-Man*, *Metal Gear Solid V: The Phantom Pain*, and *Ratchet and Clank: Rift Apart*. Some of these logs are quite long—

Metal Gear's logs alone are nearly seven hours—and we ask students to dip in and out of the game genres they're interested in, consuming them at random, sampling at least fifteen to thirty total minutes. We then discuss their experience in class, specifically focusing on the differences between writing traditional video game dialogue/narration as opposed to the more intimate and often experimental genre of audio logs. At this point, students are ready to learn the software.

The Audacity project kicked off in 1999, spearheaded by Carnegie Mellon professor Roger Dannenberg—a specialist in a field he calls "computer music"—and his then grad student Dominic Mazzoni. This free and open-source editing suite can be found and downloaded at https://www.audacityteam.org/, and, unlike Twine, must be downloaded to function. Launching Audacity shuttles students into a complicated-looking screen with a dizzying number of options. However, students need to learn very little to use Audacity for our purposes, although we'd still recommend demonstrating the program to them over a full class session. We first point out the large red circle near the top of the Audacity program. This is the record button. Press that, and your laptop will immediately begin recording, your voice dipping up and down in real time. For most students, that's all they need to learn. Hit the red record button, perform or direct your audio log, hit the stop button, and then choose "Export" from the "File" menu. That will spit out a .WAV or .MP3 file that just about anyone can listen to. Much like our simple code—[[what the player sees|where the player goes]]—in Twine, this is enough for most basic functions (Figure 14).

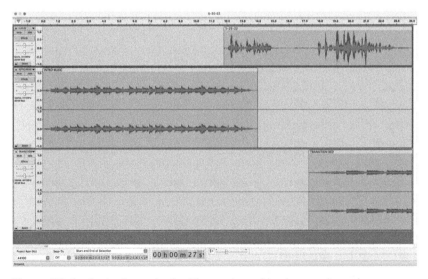

Figure 14 An Audacity project with vocals and background music.

However, there's a lot more to play with. For students without a laptop, we recommend they record audio on their cell phones and then use the "Import" command right below "Export" to begin editing. For students without a laptop or cell phone, we recommend borrowing a computer from your university library if that's an option or booking your audio log class days in a computer lab. If a student needs to edit their audio, all they have to do is highlight the audio they want to remove and hit delete on their keyboard. To add background music or sound effects, just find a track you like and return to the "Import" command mentioned earlier. There's also a rich menu of options under the "Effect" tab that allows students to add distortion, reduce noise, and even fade in or out. If the student has a microphone, all they have to do is plug it in before starting Audacity, and the program will automatically pick it up. Let students explore Audacity for even a half hour, and they'll quickly figure out the basic functionality needed to record high-quality audio logs.

Once the students have been introduced to the software, we stage a low-stakes activity to ensure that students can effectively use Audacity—while also encouraging collaboration. See the "Centurion Ark" assignment in Chapter 13. We then introduce the high-stakes assignment you'll find next. In terms of pedagogical scaffolding, this assignment is similar to our earlier Twine project, but it also jettisons some of the restrictions while also encouraging students to collaborate. We recommend Audacity for writers interested in podcasts or radio dramas who also want to direct actors. It's a perfect complementary skill to learn that can then be incorporated in Twine, Ren'Py, or many other video game design suites.

High-Stakes Audacity Assignment

Level 1: Assignment

As we've seen in games like *The Stanley Parable*, *BioShock*, *Ratchet and Clank: Rift Apart*, *Spider-Man*, *Everything*, and *Metal Gear Solid V: The Phantom Pain*, audio logs can have a positive or negative effect on story in video games. Typically, mainstream games journalism sites like *Giant Bomb* define audio logs as "voice recordings that are left behind by former inhabitants of the area the player is exploring," and you might have seen examples of them in all kinds of different games.

Now that you're in the process of learning Audacity, it's time to practice this technique for ourselves. **For this assignment, you will produce five two-minute audio logs for a fictional video game of your own creation in addition** to a written narrative pitch explaining your game and the context in which the audio logs will be utilized. Your five audio logs do not need to convey the narrative of the entire game—although they may—but you should explain in your narrative pitch how these logs function in the game. Who is speaking? Where does the player find the logs? What are you attempting to convey?

For this assignment, we're casting off one of the narrative restraints from game jam #1. Now, you must mix and match only one genre and one theme from the table.

Note: Adding music or sound effects to your audio logs is not required but encouraged.

Level 2: Student Teams and Actors

At least once during this semester, you will be required to participate in a team with another student or group of students for a high-stakes assignment. You can form a team on your own—no one outside of this class, please—or you may ask us to assign you a partner. But what does a team look like in our class?

First off, you will receive separate grades. Each student will be responsible for creating one act of their game. For example, Rico and Tiziana decide to form a team. Rico is responsible for producing the five audio logs and narrative pitch for Act One, introducing the world and characters and leaving a bunch of threads for Tiziana to pick up on. Tiziana, meanwhile, produces five audio logs and a narrative pitch for Act Two, finishing the story. Obviously, Rico and Tiziana should chat about their world and their characters and their narrative, but their grades will be completely separate. If three students form a team, then you will produce three acts. If four, then four. **No more than four students to a team, please.**

Additionally, you are not required to read and perform your audio logs. **We encourage you to find someone who will record these logs for you. However, if you are going to perform your audio logs yourself, you must perform them.** Do not read them in a monotone voice. Perform them. If you're uncomfortable with this, we encourage you to find actors outside of our class.

Just as we noted, we offer the student collaboration path for every high-stakes assignment we've included. However, we won't repeat that text moving forward and will instead only include unique instructions.

A few further resources on Audacity:

https://forum.audacityteam.org/index.php
https://youtu.be/vlzOb4OLj94

https://mixkit.co/free-sound-effects/game/
https://pixabay.com/music/

Ren'Py

In Chapter 7, we covered the history of the visual novel, beginning with its origins in Japan and culminating with its spread across the globe. As a reminder, visual novels are electronic choose-your-own-adventure novels. Nearly all of them feature art, usually a background image and a character drawing, and most employ branching narratives, choices, and music. Many of the most popular visual novels have been created in Ren'Py—including the aforementioned *Butterfly Soup*. Although visual novels were a niche genre in the West in the 1980s and 1990s, they've since exploded in popularity. Steam—the major online distributor for video games—tracks how many people are playing any one genre at any particular time, and, as of this writing, there are currently 7.2 million people playing a visual novel. Learning Ren'Py is your first step into that larger world. For many writers, it's their first foray into an engine specifically designed for releasing video games commercially.

Launched for free as a open-source web project by Tom "PyTom" Rothamel in 2004, Ren'Py is a surprisingly complicated tool for building visual novels. The name is a play on the Japanese term *ren'ai*—or "romantic love"—and Python—the coding language that Ren'Py uses under the hood. Unlike Twine where all coding is optional save for our basic [[what the player sees|where the player goes]] code, Ren'Py requires the user to learn a basic understanding of Python—including strings and more—to produce something that is even halfway playable. While a writer with no experience with coding can export a simple Twine game in perhaps fifteen minutes, Ren'Py requires significantly more time and dedication. However, a game produced in Ren'Py often looks much more professional, and, unlike most Twine games, Ren'Py games are often sold commercially. The decision between using Twine and Ren'Py often comes down to your intended result. Do you want to release your text game as a free-to-play art experiment, or do you hope to sell it to players on Steam? The choice of using Twine or Ren'Py in our classrooms boils down to whether or not we as professors want to focus on teaching video game writing skills or programming/bug testing. You must also factor in how much class time you want to spend learning a program. Most students can figure out Twine during a brief in-class demonstration. Ren'Py may take weeks.

Luckily, learning Ren'Py is in some ways more straightforward than learning Twine or Audacity. Unlike those relatively simpler programs, Ren'Py includes its own tutorial masquerading as a visual novel. After installing the program from https://www.renpy.org/, writers will find a robust guide that explains everything from the very basics—how to get Ren'Py to render text onscreen—to the most advanced concerns—how to translate your game into different languages or how to incorporate mini-games. However, this tutorial is lengthy, complicated, and occasionally overwhelming for students. Instead of spending a great number of pages here running through the Ren'Py's basics, we strongly recommend that you simply play through the tutorial. When we've taught Ren'Py, we've assigned the tutorial as homework and then coupled that with an in-class demonstration (Figure 15).

Figure 15 Ren'Py fortunately provides its own built-in tutorial.

For examples of compelling visual novels, we immediately steer students to the aforementioned *Butterfly Soup*. But it's also worth noting *Coming Out on Top*—a gay dating sim—*Doki Doki Literature Club!*—a meta horror affair in the *NieR: Automata/Stanley Parable* vein—and *Analogue: A Hate Story*—a transhuman meditation set in the far future. Because these games are typically created by only a handful of people—often a single writer, composer, and artist—and don't require the bloated budgets of AAA video games, they can delve into issues that massive corporations would likely be reluctant to touch. Often, that results in games speaking to communities that are often ignored by franchises like Call of Duty and Madden. For writers who want to make the leap from hobbyist into full-time developer—or anyone looking to pad out their work portfolio with well-written games that also demonstrate their ability to code—Ren'Py is an obvious choice.

Next, you'll find our high-stakes assignment for learning Ren'Py. Typically, we scaffold toward it with the low-stakes Twine assignment from earlier, only we swap the Western genre for whatever the class seems inclined toward—say sci-fi or fantasy or literary fiction. We should note here that although we're covering four development tools in this chapter, we absolutely do not recommend teaching four development tools in a single semester to undergraduates. This always results in a choice—are we going to sacrifice Ren'Py or Bitsy, the two most complicated tools of the set. For us, we usually drop Ren'Py simply because the writing of a visual novel too closely matches the writing of a Twine game. Both prioritize branching narratives and static images and can be slightly redundant for students. However, Ren'Py remains the superior option for students/writers who want to join the game industry as professionals, and, for that reason, it's an important tool to consider as you plan your classes/which tools you want to learn as an artist.

High-Stakes Ren'Py Assignment

Level 1: Assignment

Now that you're in the process of learning Ren'Py, it's time to create a fully fledged visual novel. **You will produce a visual novel with a minimum of 3,000 words and at least four branching paths and endings. We want to see new and fresh writing throughout, and we want to see you provide the player with meaningful choices.**

For this assignment, we're casting off some of the narrative restraints from our previous assignments. Now, you can make a game about anything you want, as long as you select one of the themes from the table.

A few further resources on Ren'Py:

https://www.renpy.org/doc/html/quickstart.html
https://youtu.be/_-hNdKUygxE

Bitsy

While the first three development tools we've focused on work wonderfully with text and even audio, none of them easily support moving graphics. That all changes with Bitsy. Released as a free web tool by Adam LeDoux in 2017, Bitsy allows writers to create simple 2D adventures that resemble the primitive graphics of an Atari or Nintendo game. Bitsy games take place in 16 x 16 tile rooms. Each of a room's 256 blocks are comprised of 8 x 8 tile drawings. Each drawing can have two frames of animation. Both the simplicity and sheer lack of options make it easy for writers with little to no artistic experience to draw cartoony renderings of everything from meowing kittens to knights brandishing swords. But it also allows more experienced artists the freedom to chain tiles together into elaborate drawings far larger than the limited scope of an 8 x 8 grid. There's no easy way to initiate combat in a Bitsy game, but the player is typically free to explore a series of linked environments, talking to other characters, picking up items, and even making decisions that will lead them down branching paths. Bitsy takes all of the skills writers and students learn in Twine, Audacity, and Ren'Py and utilizes them in a game that finally features graphics.

Much like the previous development tools, we ask our students to play a copious amount of Bitsy games before we draw even a single tile. We begin by assigning Claire Morwood's clever *Tutorial* (https://www.shimmerwitch .space/bitsyTutorial.html), a Bitsy instruction manual that doubles as a game. We then introduce the below list of Bitsy games and ask students to play through a handful. However, it's just as easy to link to itch.io, click on the Bitsy tab, and select an assortment of games filtered by topics like queer or adventure or cats (https://itch.io/games/tag-bitsy, Figure 16).

A Night Train to the Forest Zone
W.I.P
Rain
Vitreous
Going by Train to the Human Palace
The Last Days of Our Castle
Tome Full of Stars
Under a Star Called Sun
Penroses
Metrodungeon
Limbo Train

Waking Up While the Sun Sets
Midnight Bakery

Figure 16 One of the opening scenes from the Bitsy game, *The Last Days of Our Castle.*

After playing a sampling of these games, students quickly deduce that Bitsy is best equipped to produce small, thoughtful experiences that de-emphasize combat as much as possible. A compelling Bitsy game might focus on nostalgia or PTSD or homesickness or regret. It's the perfect vehicle for writers attempting to craft interesting games that push back against the dominant AAA trend in games where the first reaction to anything is violence. In our experience, students only struggle while using Bitsy if they try to twist it to feature a great deal of combat. Writers with programming experience can use variables and even modify the code in the game data to produce startling results, but even this can only push Bitsy so far.

We recommend following Claire Morwood's *Tutorial* for an in-depth look at how to use Bitsy, but essentially you only need to learn four tools inside the Bitsy program—accessible online at www.Bitsy.org. The Paint window is essential. Here, you can draw four kinds of objects—the Avatar, Tiles, Sprites, and Items. The Avatar is the playable character, and there can be only one. You can, however, draw as many Tiles, Sprites, and Items as you wish. You draw each of these objects in the aforementioned 8 x 8 grid, clicking each box to switch between two colors—red and blue, for example. Each object can be animated, and you can type in dialogue below each Sprite—a non-playable character or a physical object like a plant—or an Item the player can retrieve. To the left of the Paint window, you'll find the Colors tool. Here, you can select three colors you can use in your game—one for backgrounds, tiles, and sprites. To the left of Colors is the Room editor. Here is where you design each room in your game, laying in tiles and sprites. Finally, there's the Exits and Endings tools. This allows you to link one room to another—laying in exits and entrances—and also to mark the ending condition of your game. This is all you need to begin crafting thoughtful Bitsy games of your own (Figure 17).

Needless to say, Bitsy is significantly more complicated and difficult to learn than something like Twine or Audacity. However, in our experience, it never takes more than two dedicated class sessions for students to pick up the various nuances that go into making a Bitsy game. This time commitment always proves worth it as students turn in playable stories that truly feel like video games in a way Twine and Audacity projects sometimes don't. You can't just turn in a literary short story with a choice or two in Bitsy. You must wrestle with what it means to write a game. Additionally, the intense limitations Bitsy puts on the writer—in terms of the miniature sprites and the severely limited number of available colors—really force students/writers to get creative. If students struggle under these restrictions, there are a number of community mods that allow Bitsy writers to add all sorts of advanced features. We almost always end our courses with the following Bitsy assignment, but sometimes we've swapped that out for the Ren'Py

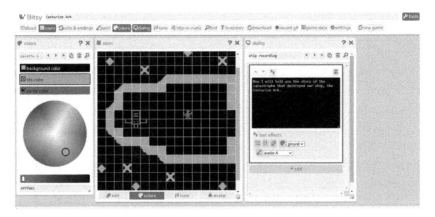

Figure 17 A hypothetical student designs a Bitsy game based off of a classroom prompt.

assignment. Again, we'd strongly urge you not to try to teach all four programs in one semester. Three is plenty—and perhaps that number is even too high—and we'd urge you to pick between Ren'Py and Bitsy at the end because they're the most professional and complicated to learn.

High-Stakes Bitsy Assignment

Level 1: Assignment

So far this semester we've worked with words. We've focused on meaningful choices in Twine, and we've honed our dialogue in Audacity. But what about graphics? What can you accomplish as a solo developer when you have all the tools of game design at your disposal—narrative, graphics, and player interaction?

Now that you're in the process of learning Bitsy, it's time to create a fully fledged game. **For this assignment, you will produce a Bitsy game with a minimum of 20 rooms and 1,800 words**. You will be tasked with designing every room and character, learning the surprisingly powerful Bitsy software, and crafting a game that has reflected everything you've learned this semester.

For this assignment, we're casting off some of the narrative restraints from the two previous high-stakes assignments. Now, you can make a game about anything you want, as long as you select one of the themes from the table.

A few further resources on Bitsy:

https://bitsy.fandom.com/wiki/Bitsy_hacks_collection_on_GitHub
https://bitsy.fandom.com/wiki/Tutorials

Additional game engines/creative writing tools:

RPG Maker MV
GameMaker
Godot
Stencyl
Unity
Unreal

Breaking into the Industry

As we've made clear time and again, this book is not meant as a guide for breaking into the video games industry. Instead, it's designed to inspire writers, teachers, and students to view video games as artistic projects that they can create. This medium is for everyone, and the barrier to entry—in terms of both creating and releasing games—is lower than it's ever been, thanks to a number of simple tools and websites that grows each and every year. However, much of what we've discussed throughout this book can be used toward finding a job in the industry. In this section, we'll cover some of those options and paths.

To begin, it's useful to define and explore a few terms. The first term our students usually ask us about is "narrative designer." They might casually stumble upon a job listing seeking a narrative designer and want to know if they're a fit for the position based on what they've learned in our classes—or perhaps from reading this book. The answer is a complicated yes and no. Some video game companies use the term "game writer," "writer," and "narrative designer" interchangeably. Many do not. For many game studios, the narrative designer is the person who literally designs the structure that relays the narrative to players—for example, the text boxes of a role-playing game. In that situation, the narrative designer might never write anything resembling traditional creative writing. However, some studios view the narrative designer as someone who both writes the game narrative and designs the structure of the said narrative—how many choices and branching paths to include and so on. Unfortunately, there's not a lot of standardization around the term "narrative designer," and we'd encourage you to research any narrative design positions as much as possible before applying. Writer or game writer, on the other hand, is much more straightforward, requiring the kind of skills we've developed throughout this book.

Another confusing element of the game industry is the difference between "solo," "indie," and "AAA" developers. As a term, the solo developer is fairly straightforward, applied to someone who creates a game completely on their own. Many of the Twine and Bitsy games we've discussed in this book are the result of solo developers, but there are also mainstream commercial examples—*Undertale*, *Cave Story*, *Stardew Valley*, and *Papers, Please*. For obvious reasons, these games are typically small in scope and usually stick to 2D graphics rather than labor-intensive 3D. We recommend the solo development route to people who want total control of their projects and can also draw, code, and write music.

Indie developers then are usually small teams of people working outside the massive corporate structures of traditional AAA game development. Usually, these teams feature a half-dozen or sometimes nearly a hundred people, but they lack the resources of enormous developers like EA or Ubisoft. An indie game like *Shovel Knight*, for example, was developed and self-published by five people after crowdfunding resources through Kickstarter. A game like *Dead Cells*, on the other hand, was developed by eight people and then published by Playdigious, a company focused on bringing indie games to market. While many indie games are simplistic and 2D like their solo developed brethren, there are many examples with expensive-looking 3D graphics as well—*Firewatch*, *Subnautica*, and *Octodad: Dadliest Catch*. The benefits to working in the indie space are twofold. (1) The developers tend to have fewer restraints creatively and can write games about sensitive or provocative material that maybe wouldn't flourish in a corporate environment—*Cibele*'s frank depiction of female desire or the anticapitalist rhetoric of *Disco Elysium*—and (2) because indie teams are dramatically smaller than their AAA counterparts, that often means developers share in the profits of their games. Although there are examples of indie developers becoming rich seemingly overnight based on the release of a single game—Team Meat and Super Meat Boy, for example—this is the extreme outlier. We've been part of indie teams where there was no profit sharing, and we've been part of indie teams where the finished game barely sold any copies and no one made a dime. There's a risk/reward element inherent to being part of an indie team, and that's why we recommend this work to folks who want a higher degree of creative control but can also stomach the anxiety/financial weight of being part of a transient team that may not exist five years down the line.

AAA development is often the steadiest work one can find in the games industry. These are corporate jobs that guarantee paychecks, but there's generally very little profit sharing except for the most high-ranking positions. As for an AAA team's scale, this can vary wildly. Bethesda—one of the major developers in the industry—launched *Fallout 4* in 2015 off the backs of a little more than 100 developers. *The Last of Us: Part II*—released by Naughty Dog in 2020—relied on a 2,000-person team. Generally, one drawback to working in the AAA game space is the lack of creative freedom. Unless you're near the top of the command chain—like writer/director Neil Druckman at Naughty Dog—it's difficult to have much impact on the overall story, tone, or scope of any given project. Likewise, AAA games don't have the freedom to tell the kinds of provocative or personal stories you often see in indie or solo games. This is beginning to change—*Red Dead Redemption II* attempts to infuse its pulpy Western tale with meditations on both illness and human nature that wouldn't feel out of place in a Cormac McCarthy novel, while the aforementioned *The Last of Us* series aspires to the apocalyptic melancholy of Colson Whitehead's Zone One—but generally these games exist to entertain and empower players more than to move them on any deeper, thematic level. What nuanced emotional resonance is there to be found in something like 2018's *Shadow of the Tomb Raider*—which refuses to interrogate its colonial underpinnings—or something nostalgic and straightforward like *Super Mario Odyssey*? Low to mid-tier AAA developers sacrifice creative control for job stability, but even that's not a given in this rapidly changing industry where turnover is both frequent and constant. But if you want to play on the absolute biggest stage with the best-looking graphics, then the AAA path may be for you.

Compounding the difficulty of working in the games industry is the persistence of crunch, the lack of any widespread unionization, and the continued harassment of women and marginalized peoples both at work and in online spaces. All of these topics could and have supported individual books, but we'll cover the basics here. Crunch is industry jargon for when developers are asked/forced to work overtime to ensure their project releases on time. According to *The Washington Post*,

> [In] a 2019 survey from the International Game Developers Association, 40 percent of game developers reported working crunch time at least once over the course of the previous year. For the majority of these developers, crunch wasn't just a few extra hours or a long weekend, but at least 20 extra hours on top of their standard 40-hour workweek. Just 8 percent said they received extra pay for those hours. (Thomsen 2019)

This exploitative workload is a direct contributor to burnout and is perhaps only possible in an industry where unionizing has been historically discouraged and blocked at nearly every corner. Even this, thankfully, is slowly changing, and in 2022 Raven Software, a developer of Call of Duty under the control of Activision Blizzard, voted to certify the Game Workers Alliance, the first union at an AAA American studio. Harder to rectify, however, is the continued harassment of women and marginalized peoples in the game space. In 2020, waves of stories of prolonged harassment in the wake of the #MeToo movement rocked the games industry (MacDonald) accusing high-profile individuals at companies like Ubisoft and Riot, among many others. This is unfortunately inexorably linked to the #GamerGate movement of 2014, where various women in the industry were aggressively harassed either for critiquing sexism in games—Anita Sarkeesian—or for simply being a female developer online—Zoë Quinn.

Graduate school is also a valid path for folks, and talented developers— like the aforementioned Nina Freeman behind games like *Cibele, Kimmy,* and *Freshman Year*—matriculated from the prestigious NYU Game Center program. However, most of these kinds of programs focus on overall game design and not wholly on game writing/narrative design. Likewise, most traditional MFA programs in creative writing are resistant to game writing or interactive fiction—although this is in the process of changing.

If, after reading and reflecting on all of the above, you still hope to enter the video games industry, where does one begin? This is the moment where we steer our students toward building a work portfolio. In our classes, we instruct our students to publish every high-stakes project to either itch.io or Soundcloud, ensuring they have a fully developed work portfolio by semester's end—of course, we also give them the option to lock and password protect their projects for privacy's sake. But let's say you haven't finished any of the high-stakes assignments we've included earlier in this chapter or perhaps you just want to complete a few more quality games before releasing them online. In that situation, we steer writers toward game jams.

A "game jam" is a term for a semiformal event in which hobbyist or amateur game developers form small teams and quickly create a game under a specified theme. During our classes, we often show students a repository (https://itch.io/jams) listing over 100 concurrent game jams as of this writing. Themes are extremely varied. Some prioritize content like summer, or queer, or mecha games, while others focus on specific development tools or even parts of the world. The vast majority of these Games Jams are done online, but there are physical counterparts as well. If you'd rather work in-person, check to see if there's a chapter of the aforementioned International Game Developers Association (IGDA) in your nearest city. Once you join a specific game jam, you're often invited to a Discord channel with all the other participants. Some people assemble their teams ahead of time, but most folks enter the Discord and list what they bring to the table—perhaps writing or music or coding or art. People build teams from there and break off to design their games. Joining a game jam has its pros and cons. Because this is all unpaid labor, it's often difficult to ensure that everyone follows through with their responsibilities and finishes the game on time. However, even in that case, writers should come away with work for their portfolio in the form of scripts or whatever writing they accomplished for the unreleased game. As for the pros, game jams ensure that an audience will play your game. Usually, these events culminate with people voting on the best finished game, and Twitch streamers who focus on game jams will often livestream themselves playing all the entries from the game jam themes they're interested in. There have also been many examples of commercial games that began as game jam projects—*Goat Surgeon* and *Super Hot*, to name only two. In our classes, we call each high-stakes assignment a game jam to hopefully persuade our students to try their hands at a real game jam at semester's end.

So, what do you do after you have a strong collection of completed games and/or scripts? At this point, it's time to build your website and work portfolio. Luckily, this process is easier than ever. As we wrote earlier, our students publish their games directly to itch.io. This is an online repository for solo and indie games that doesn't require a complicated approval process like on Steam. Cruise over to Itch, and you'll find thousands of tools and games, some for sale and some for free. Everything from simplistic Bitsy and Twine games to console games like *Night in the Woods* and *Celeste* are published on Itch. We demonstrate to our students how to publish games on Itch, but the process couldn't be easier. If you can upload a picture to social media, then you can upload the finished file that Bitsy or Twine or Ren'Py spit out. It's the same process, and students pick it up within seconds. Conveniently, Itch will store all of your completed games on a single page. This, for most of us, is our work portfolio, the link we'll share with developers when we apply for jobs in the industry. Occasionally, you'll want to fine-tune your application, sending the hiring manager the link to just a single game or file, but having all of these materials in one easy-to-navigate online repository is absolutely invaluable. Plus, you never know who will play—or buy—your games when you're not actively applying for jobs. Although most job ads will ask to see finished games along with your resume and cover letter, some will instead ask for scripts or even short stories. In that situation, you can simply attach the requested Word document to the email application. However, some game writers do include these materials in their online portfolios. That decision is up to you.

Although Itch is the easiest method for both releasing your games online and building an easily accessible work portfolio, some game writers build their own websites with dedicated portfolios instead. We'd only recommend this to writers with a strong handle on html. Alternatively, a middle ground would be building your own website on something like WordPress and linking to your games on Itch. Resumes for the games industry should focus both on your writing/game experience and on the programs you've learned. If you're skilled at Excel or Twine, don't hesitate to mention that. The same is true for any programming languages you know or if you're a credible artist or composer. A jack of all trades is especially useful on a small indie team where each individual needs to wear multiple hats, unlike the AAA space where nearly every task is extremely specialized. A strong job letter addresses the needs of the ad while also explaining your experience and why you want this particular position. We always tell our students to tailor everything to the job ad. If the position is asking for a sci-fi writer, mention some sci-fi games in the letter and try as much as possible to steer them toward your own sci-fi games or writing.

But where do you find any of these mysterious game jobs? This is ever changing—especially in a cutting-edge industry like games—but the most consistent source comes from various job boards. https://gamejobs.co/ is the largest job board in the industry, focusing primarily on AAA jobs. As of this writing, the search term "writer" generates over 200 open positions, everywhere from New York City to Warsaw to the rapidly growing remote workplace. Another quality board, https://hitmarker.net/, currently lists forty. A job board specializing in indie jobs, https://www.workwithindies.com/, lists seven. Surprisingly, LinkedIn remains an excellent way to find jobs in the industry, and we always recommend to our students interested in working in games to build a profile and to set an alert for any game writing jobs. Discord—a messaging app popular with Gen Z and younger millennials—is also a wonderful resource for job hunters. The IGDA SIG Game Writing Channel has a job postings forum that lists intriguing positions that often fall off the radar of the larger job boards. It's also not uncommon for game writers to directly reach out to small indie teams to see if they need any writing. Game writers will often lurk on Kickstarter, searching for video games that have already been funded or are about to be. Newly flush with cash, these indie teams may be more willing to take on a paid writer after a successful Kickstarter campaign. Twitter—despite its penchant for both trolls and harassment—remains a viable means to find employment in the industry. Game writers are certainly not above direct messaging accounts for small-scale games early in development to see if they need any writing. We've done this ourselves—and landed jobs this way—and highly recommend scanning the #ScreenshotSaturday hashtag for a seemingly endless list of indie games currently in development.

Overview

Now that you've honed your creative writing craft and thought through the various issues introduced by our game design chapters, it's time to begin writing games yourself. That means learning software, including Twine,

Audacity, Ren'Py, and Bitsy—each better suited toward different types of projects. For readers interested in joining the video games industry as professionals, we've laid out a number of important terms—"indie," "AAA," "narrative designer," and so on—and the problems of crunch, workplace harassment, and sustained online harassment that still plague this industry. For the committed, we recommend scanning game jams, job boards, and various online spaces like Discord, LinkedIn, and Twitter.

13

Peer Workshops and Collaborative Writing

What *Is* Workshopping?

Sign up for any creative writing class, and you'll likely be presented with the following scenario. Students read a few short examples of fiction, creative nonfiction, and/or poetry before discussing *how* the pieces work rather than *what they mean*. Maybe the class speaks with a visiting writer or follows the instructor through a series of guided exercises. But generally, the centerpiece of any creative writing class is the workshop, where students submit packets of twelve to twenty-five pages of creative work to the rest of the class before spending approximately forty-five minutes discussing how the writer might best revise their piece/s. Often, the writer sits silently taking notes while the class and instructor ponder aesthetic, albeit puzzling, questions like "Does this work?" or "Does this character behave in a believable way?" This is sometimes referred to as the traditional workshop model and has served as the cornerstone for many creative writing programs since it was first popularized by the Iowa Writers' Workshop at the University of Iowa in mid-century.

Much has been written about why this narrow model disadvantages or even discriminates against wide swaths of creative writing students. Matthew Salesses in *Craft in the Real World: Rethinking Fiction Writing and Workshopping* writes,

> I sat silently in an MFA workshop while mostly white writers discussed my race. I had decided not to name the race of any character, Asian American or otherwise—but the workshop demanded that the story inform "the reader" if my characters were like me, people of color. A common assumption lies behind this phenomenon: that no mention of race is supposed to mean a character is white. I didn't have to ask why the white writers in the room never identified the race of their white characters. I already knew why: they believed that white is literature's default. I just couldn't say so. (Salesses XIV)

Felicia Rose Chavez in *The Anti-Racist Writing Workshop: How to Decolonize the Creative Writing Classroom* agrees, writing, "Per the pedagogical rite of passage, the writer is forbidden to speak. This silencing, particularly for writers of color, is especially destructive in institutions that routinely disregard the lived experiences of people who are not white. This matrix of silence is so profound it enlists writers of color to eradicate ourselves" (Chavez 3). Both of these writers/instructors make convincing cases for decentering the traditional workshop and instead refocusing on student agency, allowing emerging writers to take control over their own development and speak to whatever communities and traditions they find most appealing and central to their identities. Much of our work in this chapter owes a debt to Salesses and Chavez and builds from their hard work, not to mention the dedication of so many instructors evolving what a creative writing class is and does in the twenty-first century.

So then, what does it even mean to workshop writing for video games, and what are the benefits? Collaboration and receiving feedback are two absolutely vital skills for producing video games at any professional or amateur level. Nearly every game you've ever played relies on collaboration or feedback—even games produced by a single human being like Toby Fox's *Undertale*. Learning how to receive and give feedback—and also how to collaborate with other artists—is a soft skill that all too often is ignored by folks hoping to write games—from self-published visual novels to AAA open-world console releases. These workshop models simulate those aspects, but because an open-ended or nonlinear video game is so different from a set twelve-page prose short story, traditional workshopping might not even benefit you, your students, or your project. We encourage flexibility throughout this chapter and would steer folks toward the rapid peer review option if they want peer feedback without the in-depth preparation and hassle of something resembling a more thorough workshop. This chapter will not only lay out three different workshop models with their pros and cons but will also encourage you to identify and select the model that will most benefit your project. Agency is one of the driving factors of engagement on any college campus in this country. It's time for you to take control and decide for yourself.

Rapid Peer Review

Sometimes a writer doesn't benefit from sharing their work with ten to twenty of their peers. This kind of detailed feedback—whether it involves a traditional workshop component or not—can be rather time-consuming and usually works best when the project up for discussion—be it a video game or a short story or a series of poems—is long enough and revised enough for the reviewers to really sink their teeth into. But what if a writer desires feedback on a shorter experience like a brief Bitsy adventure or a Twine journal entry? What if a writer really needs rapid feedback that they can immediately apply to another more substantial draft?

In the previous scenarios, rapid peer review might be the best option. Using this method, groups of only two or three students are linked together—either via self-selection or chosen by their professor—and given time either outside of class or during class to play their partners' prototype games before completing a peer review report. Of the three workshopping options we're exploring here, this is the least time-consuming option for both students and professors. Instead of dragging workshops out for multiple weeks as students play and provide feedback for every game produced by the many members of the class, workshopping is instead limited to one or two class sessions where all students receive swift and targeted feedback. This option, in our experience, works best in courses with large class sizes or courses that devote large swaths of in-class time to learning specific game design suites or game studies theory. Likewise, it works well for certain general education students or first or second-year students who might not be ready to take on the responsibility of directing a class session or helming multiple revisions. Although rapid peer review can be used as a prelude to a lengthier workshop, it primarily attempts to help students identify a few isolated opportunities for revision without being overwhelmed by the litany of issues that may arise during a meatier workshop.

Perhaps the most obvious drawback with the rapid peer review method is that students only receive feedback from one or at most two other students. I'm sure we've all been in a class or run one ourselves where we attempted a similar peer review assignment on a first draft of an academic paper. If the assignment isn't phrased in exactly the right way, students may default to pointing out what's already working so as not to offend their peers or "give them extra work." Below, you'll find our rapid peer review assignment which tries to guide students as much as possible toward providing quality feedback that will help any writer toward an improved draft. In our classes, we embed this rapid peer review component into our larger formal game writing assignments—which you'll see later on. The goal then is for emerging writers to begin viewing feedback as a necessary component of all writing—not as "extra work" they're dumping onto a peer.

Rapid Peer Review Assignment

After everyone has submitted their Twine game, I will assign each student to a small two or three-person group. If you'd instead prefer to work with someone in class who you know and trust, simply let us know.

You must play and complete your peer's game before class time next week. Before the end of that day's session, you must write a 500-word Peer Review that will help your partner revise their game. Submit this on our Discussion Board.

First, very briefly **describe** what a student's game is about. This should be the shortest section. Second, tell us briefly about the game's **strengths**. Finally, finish with **prescription**, a paragraph explaining very specific elements of the game that still need work and ways in which your peer can revise their game for the better by the end of the course. This should be the longest section.

The description-strengths-prescription model is something we'll discuss again in the "Traditional Workshop" section. Essentially, it's a modification of the traditional workshop that prevents students from either writing only praise or writing only criticism. Sal first experienced this model in a workshop steered by Cathy Day and was floored by how it broke down both the traditional workshop into manageable and more equitable blocks. Initially, students may push back against the description component of this response. "What's the point?," they sometimes ask us. But soon enough, one of them will write a game with a certain aesthetic intention that the rest of the class doesn't yet see. The sometimes vast gulf between their intentions and their peers' descriptions of their work will help streamline what they should focus on in a revision.

Traditional Workshop

Sometimes rapid peer review isn't enough. Sometimes a writer needs input not only from one or two peers but from an entire classroom including their professor. In those cases—typically when students are writing longer, more elaborate games later in the semester that have already gone through a revision or two or in upper-division courses in general—the traditional workshop or the alternative critical response process are superior options to the rapid peer review.

Most folks who have taken a creative writing class are probably familiar with the traditional workshop. Students submit their projects for review to the entire class, and then forty-five minutes of class time is dedicated to the writer's peers and professor providing artistic feedback aimed toward a revision. Although there are many drawbacks and concerns with using the traditional workshop—see the commentary by Chavez and Salesses—we understand that the traditional workshop remains the dominant form of creative writing pedagogy in the country. For those instructors determined to employ a traditional workshop in their courses—and also for students who prefer them—we're including our own modified traditional workshop here that builds off of Cathy Day's description-strengths-prescription model outlined earlier tailored specifically for a game writing classroom that intends to minimize the many drawbacks of this admittedly retro pedagogical tool. Below, you'll find an excerpt from the assignment sheet we often use in game writing classes. Please note that we always allow students to choose— do they want to go with the traditional model, or would they prefer the critical response process?

Workshop Assignment

Our primary in-class activity this semester will be workshopping. But what does it mean to workshop? In short, each student will submit a Twine or Bitsy game twice a semester resulting in a portfolio of completed video games at semester's end.

You will play all of your peers' games and be expected to post a 600 to 1,000-word critique using the method detailed below on our Discussion Board. During class, you'll be expected to provide oral feedback that guides the writer toward a stronger revision.

When it's your turn to submit a game, you will choose from two workshop models. As a class, we'll workshop two to four students each week.

(1)   The Traditional Model

In this model, the instructor will facilitate a 45-minute workshop by guiding the discussion and asking for feedback from the rest of the class. The instructor will ask the class to **describe** what the student writer's game is about. Then, the instructor will ask the class to describe the game's **strengths**. Then, the instructor will ask the class to provide **prescription**, that is, tangible methods to sharpen the creative work in a revision. The instructor and students will lay out a very clear revision plan for the student writer. Finally, the instructor will end the session by asking the student writer if they have any additional questions about their game for the rest of the class.

This option usually works best for students who aren't completely sure what to work on in their revision. It's great for students who want to get lots of revision-oriented feedback aimed at their final portfolios for the end of the course.

What's Expected of You During Workshop as a Student Reviewer

When it's not your turn to submit creative work to the class, you will be a student reviewer, responding to your peers' games and giving them verbal and written feedback. For other students' workshops, you must play their games and then post a 600 to 1,000-word response on our Discussion Board at midnight the day before class. These responses are not meant to be pats on the back, nor are they meant to be mean-spirited takedowns. Instead, we'd like to see you follow the description-strengths-prescription method referenced earlier.

First, describe what a student's work is about. This should be the shortest section, but sometimes you as a writer will think you've accomplished one thing only to realize you've done something totally different. Second, tell us briefly about the piece's strengths. Finally, finish with prescription, a paragraph explaining very specific elements of the game that still need work. Your critiques must be posted by midnight the night before class. Please have a copy of your fellow students' games handy during class.

After your workshop, we will schedule an appointment to chat with you about your work, and I'll provide you with a critique letter. You'll receive a purely provisional grade which is tossed out when you turn in your final revisions during Finals Week.

Breaking down the traditional workshop into three segments tends to lead to the best outcomes when using this classroom strategy because it prevents students from descending into total criticism or total praise. It allows students with even the least polished projects to hear what's working, and it forces students to generate revision strategies for writers who turn in the most polished pieces. We'd also recommend including an example in the assignment sheet of a well-written student critique—with the permission of the former student in question, of course.

There are certainly benefits for students who select the traditional workshop model. In our experience, this version of the workshop works best for students who have completed larger games—something where rapid peer review might not be enough—but also for students who aren't very sure on what to work on in a revision. It's also a great tool for providing students who might not be as invested in the class or their artistic development with very tangible revision strategies and goals. Having the professor guide these sessions ensures they receive quality feedback even if they've barely fulfilled the parameters of the assignment. In that kind of situation, the professor can also make sure to focus on general artistic strategies that would benefit the rest of the class as well.

However, as outlined by Chavez and Salesses, the traditional workshop model strips student writers of their agency during the workshop. They're asked to remain silent and sit there like a fly on the wall as students and professors discuss their work. A small misunderstanding on the part of one or two students can sometimes be instructive for the student writer, but an entire class misunderstanding their work—or worse, their culture or identity—can be particularly harmful, not to mention dangerous and hostile. If students of color—or any students outside the majority—already find it difficult to speak in class, it can be extremely painful to force them to be quiet while strangers from outside their community discuss deeply personal artistic work. A successful traditional workshop—if such a thing is even possible—requires a professor who tries to meet students where they live and does not attempt to force their own ideologies—artistic or otherwise—on the class. While some folks believe that the traditional workshop is irredeemable, others like Joy Castro and Sandra Cisneros have attempted to reboot it. Castro, in her essay "Racial and Ethnic Justice in the Creative Writing Course" published in *Gulf Coast*, evokes Cisneros's Macondo Workshop:

> The Macondo Workshop itself, conceived in 1995, is the brainchild of Sandra Cisneros, whose goal was to create an environment more welcoming and nurturing than she herself had experienced as an MFA student. . . . The Macondo ground rules address various issues, encouraging compassionate mindfulness as "a practice motivated by having witnessed marginalization in our communities." My favorite passage is this: "Many of us come from places where we've been involved in long-term conflicts and have learned extremely valuable survival skills, including persistence, skepticism, and a willingness to confront others." Both of these descriptions of origins immediately work to decenter all students who have grown up in privileged environments—and to make them conscious of the fact that they have. Not many white suburban kids can identify with these descriptions, which automatically center the experience of many students of color. Next, I create a larger context for our work by consistently alluding during class discussion to literature by writers of color. There is a literary tradition, and you're invited to contribute, this strategy implicitly asserts to students of color. I use texts by writers of color as my reference points, employing examples of aesthetics, choices, and techniques from Morrison, Kingston, Erdrich, Justin Torres, and Helen Oyeyemi, for example, rather than from Faulkner, Hemingway, and Tobias Wolff. (2015) new line, right align

The drawback that the traditional workshop shares with the critical response process is simply how much time these models take up over the course of the semester. Workshopping every student in a class—let alone twice—could constitute half or even the majority of a semester's in-class time. For that reason, both of these more elaborate workshops may be better suited for upper-division courses or capstones where students already have larger projects in mind—or are ready to begin them. Otherwise, rapid peer review might be the superior option.

The Critical Response Process

In 1990, Liz Lerman—a noted choreographer—noticed something strange about artists during critique sessions of their own creative work. Instead of asking questions about how they could better their projects, they instead apologized for all the perceived shortcomings—both real and imagined—of their unfinished work. For the next thirteen years, Lerman worked toward a solution, culminating in 2003's *Critical Response Process: Getting Useful Feedback on Anything You Make, from Dance to Dessert.*

The concept behind the critical response process is simple. Instead of a professor steering a class of students through a critique of peer work while the artist in question remains silent, the artist will instead guide the discussion themselves. That requires humility on the part of the professor—they must accept a reduced position in the hierarchy of the classroom and surrender some of their power/authority—but it also requires the student writer to have a clear understanding of their own work including opportunities for improvement. The writer must spend additional time plotting out their own workshop and that requires someone who is already quite dedicated to their art. As you can already imagine, this works beautifully for students who are actively engaged in creative classes and want to produce art that is personally meaningful. It works less well for students who aren't as engaged in the class or in their own development as artists.

Although the critical response process was created in response to dance critique, it has quickly become popular in American creative writing classrooms. Below, you'll find our critical response process assignment sheet that we've found success with in both our creative writing and video game writing classes. Much of it builds from the work laid out by the aforementioned Chavez and Salesses.

(2) The Critical Response Process

In this model, the student writer will facilitate their own 45-minute workshop. First, the student asks the class and instructor for observations. What did they notice about the work? This step is designed for the student writer to hear how the workshop reacted to their game. Next, the student writer will ask a series of questions they prepped ahead of time to the class and the instructor. These should attempt to guide the student writer toward a revision. Next, the class and instructor are encouraged to ask the student writer questions. These should remain neutral and aim to get the student writer to make sense of their art. Examples include:

(a) Why do we learn that the player's father is dead at the end of the game and not closer to the beginning? How would that shift change the game?

(b) Why is the player prevented from leaving their family's home? Is there anything immediately outside worth exploring or even considering?

(c) How did the player character learn about the legend of the forsaken wizard?

If the student writer chooses, they can now ask for suggestions on revision from the class. The student writer may move at their own pace, and if they don't cover everything by the end of their workshop, that's fine. However, at the end of forty-five minutes, the instructor will call for the end of the session and ask the student writer to name a few concrete strategies they will attempt in revision.

This option works best for writers who have a very clear understanding of both their work and audience who are looking for very specific and directed feedback on isolated aspects of their work.

It's worth noting here that the student reviewer's tasks remain the same no matter which workshop model the student writer chooses. As you can imagine, many of the traditional workshop's drawbacks are addressed by the critical response process. Here, student writers aren't silenced or turned into empty vases waiting to be filled with knowledge from an authority figure—to borrow language from Paulo Freire's *Pedagogy of the Oppressed*. Instead, the student takes on the role of the teacher and steers their own artistic feedback toward whatever they believe will be most helpful in the development of their project. In this model, students are empowered and have greater agency.

However, this model requires that students are tremendously invested in their game and their own artistic development. In Felicia Rose Chavez's *Anti-Racist Writing Workshop*, she writes about asking students to fully commit to a semester-long focus on their lives as artists before even beginning workshopping. That works well in upper-division or capstone courses, but it can be difficult to mandate this dedication during intro classes with nonmajors looking to quickly fulfill core curriculum requirements. Again, we'd like to highlight the success we've enjoyed simply by allowing students the freedom to choose between these two models. This is a method we've employed repeatedly, and it was thrilling to watch students toggle between the traditional workshop and the critical response process depending on the project and their own artistic sensibilities. Students need agency to feel truly invested in a project. Don't be afraid to give it to them.

Collaborative Writing and Playtesting

Although much of this book focuses on teaching a single writer how to create and design their own personal video games that speak to audiences and communities of their choosing, we'd be remiss not to wave the banner for the joys of collaborative writing. Much more so than prose fiction, creative nonfiction, or poetry, video games are a collaborative medium relying on talents and creatives from multiple disciplines—music, art, architecture, writing, programming, just to name a very small few. Although there are modern commercial releases crafted by a singular artist and vision—the aforementioned *Undertale* by Toby Fox—and many free indie releases helmed by a single person—Anna Anthropy's experimental Twine game *Queers in Love at the End of the World* comes to mind—the vast majority of video games played by the public are assembled by teams of dozens, hundreds, and sometimes even thousands of employees. The number of writers varies from game to game—2019's *Star Wars: Fallen Order* utilized six different writers working in tandem while 2020's *Moving Out* was written by a single person—but there are far better reasons to write collaboratively than simply replicating the dominant structure of commercial gaming. The joy of collaborative writing often comes from simply pinballing your ideas, characters, and aspirations off another person, transforming them exponentially in the process.

We can attest to this from what we've witnessed in our own classrooms. Students who struggle generating dialogue or deeper emotional concerns suddenly bloom when paired up with a partner with a different skillset, background, or identity. A student who excels at worldbuilding and overarching ideas hits their stride when working with a student who favors characterization and mood. Time and again, we've seen students thrive in team settings, and we'd urge instructors reading this book to utilize at least a few team exercises in their courses. The dreaded "group work" can often elicit groans from students who've been burned by teams where they ended up taking on more than their fair share of the work. So how do we as instructors cultivate a safe space for collaboration while also engaging as many of our students as possible?

If you want your students to excel in team settings, you must first give them multiple opportunities to practice that work in-class where they can feel out the stumbling blocks and even fail in productive ways. For us, that means introducing a few simple low-stakes collaborative writing exercises multiple times before mandating any long-term high-stakes assignments. For example, take a look at the following exercise. We introduce this activity after students have familiarized themselves with audio logs—found diary entries scattered across many modern games ranging from *BioShock* to *Spider-Man*—and the necessary software—Audacity—to produce them themselves.

Low-Stakes Collaborative Writing Activity #1

Now that we've explored Audacity a tiny bit, it's time to practice using it. Imagine a 3D video game where a brave explorer is zipping across the galaxy in a brand-new top-of-the-line spaceship. Suddenly, against the velvet backdrop of Alpha Centauri, the player sees the husk of a broken-down freighter surrounded by specks of floating debris. The player decides to don a spacesuit and propels over to the broken ship. Inside, everything is dusty and silent. Puddles of green liquid collect along the floors. The controls look like something a human being might use, but the player has never seen these particular buttons or pedals before. In the control room, the player stumbles across a small orb pulsating with yellow light. The player impulsively picks it up, and a recorded voice begins speaking. It's an audio log that will play as the player continues exploring the ship.

A voice booms out, "Now I will tell you the story of the catastrophe that destroyed our ship, the Centurion Ark."

For this assignment, break up into trios. Discuss the above scenario first. What happened to the Centurion Ark? Who is speaking in the recording? What is the voice trying to convey? **And, most importantly of all, what should this message convey to the player?** As a group, **write out a three-minute audio log** beginning with "Now I will tell you the story of the catastrophe that destroyed our ship, the Centurion Ark." **When you're done, choose a member of your group to record this speech in Audacity. Try to perform it as best as you can. After recording, export the file as a .WAV and upload it to our Discussion Board.**

Generally, we give our students forty-five minutes to an hour to complete this assignment. It's important to note here that we've often underestimated how much time it will take groups of writers to finish these projects. We'd recommend always having additional time available if necessary while running group projects. Generally, however, the pressure is off with this particular assignment because they know it's a low-stakes project that counts for little more than participation points and that they won't have to work with these particular team members again—unless they choose to a few weeks later. Likewise, we make sure they begin in a playful brainstorming state, simply talking out what happened to the Centurion Ark. Instead of having to build the concept from scratch, we give them a limited canvas to work with. This helps acclimate students to the loss of total control they often have while writing solo projects, and it introduces them to adding collaboratively to a story they didn't create firsthand—the reality of most commercial writing in the video game space.

A few weeks after the Centurion Ark assignment, we introduce our students to a more complicated low-stakes collaborative exercise. First, we ask our students to play eight hours of *Skyrim* on their own, writing a reflection on the game's writing/narrative for homework. It's important to note here that *Skyrim* begins by asking the player to create their character—choosing everything from skillset, to race, to species, to gender. During class time, before we discuss their experiences with *Skyrim*, we ask them to respond to the following solo activity:

Skyrim Solo Activity

For this activity, try to imagine the character you created in *Skyrim* as an actual fictional character with a rich backstory all their own. Write the following as a post on our Discussion Board:

First, write a **character description**. What does your character look like today, eight hours into your playthrough or wherever you last left them in Tamriel? What race are they? What gender? What does their face look like? What kind of armor or weapons do they use? Most importantly, based on the way you've played, what is their personality? Do they fight for good, or are they evil? Somewhere in between? Is your character a loner or someone who always has a follower? Do you imagine your character as old or young? Have they been moved to fight the dragons assaulting Tamriel while following the main thrust of the game, or are they more interested in smaller outcomes, more playful endeavors?

Second, write a **fictional backstory**. What was your character up to before they arrived in Helgen on their way to execution alongside Ulfric Stormcloak? This doesn't need to follow any official *Skyrim* lore. Use this as an opportunity to deepen your character and explore.

Finally, write **a history of what your character has done in *Skyrim* so far**. Have they joined the rebellion? The Empire? Perhaps they've enrolled in the Bards College or joined up with the Thieves Guild? Did they acquire a house or adopt a child or get married? Fought any vampires? Maybe they escaped the dragon attack at Helgen and spent the next eight hours wandering around doing very little? Tell us all they've done that remains memorable to you.

When you're done, post this work to the Discussion Board. We'll discuss it together, and you'll use this character sheet for a group exercise later today.

This exercise fulfills a variety of needs. First, it preps students for the group work to come, but second, it focuses them to think about character and characterization, the areas we most often notice students struggling with. After they've completed the above exercise and chatted about their experiences in *Skyrim*, we then introduce them—likely during the next class session—to this low-stakes collaborative writing activity.

Low-Stakes Collaborative Writing Activity #2

For today's activity, I will break you into groups of three. Once assembled, **complete the following assignment**:

Pretend that Bethesda Game Studios, developer of *Skyrim*, has hired your small team to write a DLC (downloadable content) campaign for *Skyrim*. You don't have free rein, however, and there are a number of key concepts they are requiring you to incorporate. First, the campaign will occur on a brand-new island that players can only find by hiring a ship outside the city of Solstice. This is a brand-new area, and, unlike previous *Skyrim* adventures, this one will be multiplayer and will involve pre-designed players. **Your DLC campaign must focus on the characters you created at the start of the previous class session, and it's up to you to come up with a narrative justification for why these disparate characters must band together**. Bethesda is asking you for the following:

(1) **A description of a coastal city** where the three co-op players meet after disembarking from the ship they hired in Solstice. What does the city look like? What's its vibe? Who lives there? What is its main source of industry? Are there any lingering problems? Any preexisting factions?

(2) **An outline of a quest that unites the players you created earlier**, given to them by the ruler of your new city. What are they being asked to do, and why can they only accomplish this goal together as a unit instead of independently? **In addition to this outline, write the dialogue of the quest giver**. How does this character explain the mission to the players?

(3) **A description of two dungeons where the quest plays out**. Where are the players asked to go: underground dungeons, volcanic castles, underwater hideaways, pirate coves, mountain caverns?

What are these dungeons called, and what will players find inside? What makes them unique and different from the dungeons that *Skyrim* already utilizes? Is there anything inside these dungeons that require collaboration among players?

(4) **Descriptions of two bosses**. Where are they found in the DLC campaign? What do they look like? How do they fight? Why do they want to harm the players, and, conversely, why must the players harm the bosses? **Write their dialogue**.

(5) **Ending dialogue**. Once the final boss of the DLC is defeated and the quest is solved, what happens, and who speaks? Write that dialogue.

When you're finished, post this work to the Discussion Board. Be prepared to explain your DLC campaign in about five to ten minutes to the rest of the class.

Again, the goal here is to ease students into working collaboratively as teams while complicating what they learned/practiced in the previous assignment. Typically, we give them an entire class period to finish this work and often we give them the weekend to complete it. We also make sure that they work with different partners from the first low-stakes activity so that they've collaborated with at least four other people by this stage in the course. This is key when we move into the high-stakes long-term group projects—where students collaboratively write Twine or Bitsy games—so that students can choose team members they gel with if they so choose.

Pay attention to how the above assignment scaffolds from the previous assignment. In the Centurion Ark activity, students really only had to write a single source of dialogue—the audio log—and then worry about the Audacity software. In the *Skyrim* activity, they're asked to do a lot of brainstorming, creating narrative justifications linking their original characters, not to mention quest lines and dungeons. The low-stakes nature of the assignment, however, keeps the pressure off and allows them to play and experiment—the foremost goals of a video game writing course.

Obviously, the above two assignments aren't enough to teach students how to write collaboratively, but they are a start. For a more detailed look at a high-stakes collaborative writing assignment, please try the assignments in Chapter 12.

We would also be remiss in this chapter not to briefly discuss playtesting. In a traditional creative writing workshop, students typically read creative work in the form of packets ranging from twelve to twenty-five pages. Sometimes students will include a smorgasbord-style portfolio of poems and essays and flash fiction, and sometimes they will write stories told out of chronological order. But, at the end of the day, students still read peer work in exactly the same order, starting with the first page and moving through to the end of the submission. That isn't true with video games, so how do we both effectively playtest and peer review submissions ourselves, and teach our students how to do so?

In our experience, it's beneficial to encourage reviewers/playtesters to not simply go through a peer's game only once. Even if it's a linear experience lacking significant narrative choices, it's essential that reviewers work through the game at least a few times, testing the boundaries if not for any narrative concerns then in search of any particular bugs or issues plaguing the experience. In a game with branching paths like your typical Twine or Ren'Py game, encourage students to try and see everything—or at least as much as they're able to discover. This will give the peer reviewer a deeper understanding of the game and its scope and intentions, but if the reviewer misses some seemingly important narrative path, it also conveys to the writer/designer that perhaps this road is a little too hidden from the average player.

Playtesting can also be an opportunity for students to solicit feedback on their game from players outside the classroom community, and to choose their testers, compose the questions they will ask them, and design the approach they will take with a mind to engaging the specific audiences they have in mind for their project. If an LGBTQ+ student is working on a game intended for that community, for instance, it may be useful for them to curate a group of LGBTQ+ testers and to ask questions tailored for players with those backgrounds. Similarly, a student creating a serious game intended to teach players about a medical issue may want to test how well a general audience—one that may be unaccustomed to playing video games—is able to navigate their game. In some cases, rather than structuring a playtest around questions, designers may even ask players to record their gameplay, so they can observe common patterns in how players approach the game. Teaching students to design and implement their own playtests can help them develop a sense of agency and authority over the games they are creating and the audiences they are most hoping to reach.

It's also important to encourage your students to playtest their own games as often as possible. No matter how often we reiterate this point, we receive games plagued by bugs and broken paths every semester that make much of these student experiences unplayable. Usually, these are fixed after a quick back-and-forth troubleshooting with the student, but giving students dedicated in-class time to play through their own games hunting for bugs and moments where the writing rings false or the gameplay breaks down leads to more polished projects by semester's end. It also helps ready them to playtest their own games before releasing them to the public—if they so choose.

Overview

Giving, receiving, and incorporating feedback are essential skills that all artists must develop no matter their field. In this chapter, we've discussed how feedback has been historically deployed in the creative writing space while also interrogating the criticisms of how various workshop models prioritize white experiences and voices in the classroom. We've also provided three feedback models for game writing that we'd urge instructors and artists to adopt while also laying out exercises that scaffold toward meaningful collaborative writing. It's important to note, however, that feedback can't be provided without playtesting, a key process that all game writers must incorporate—whether they're working on their own games or reviewing a peer's.

Works Cited

Abernathy, Tom and Richard Rouse. "Death to the 3-Act Structure." Game
   Narrative Summit. *GDC Vault*. March 17–20, 2014. www.gdcvault.com/play
   /1020050/Death-to-the-Three-Act.
Adsit, Janelle. *Toward an Inclusive Creative Writing: Threshold Concepts to
   Guide the Literary Writing Curriculum*. Bloomsbury Academic, 2017.
Alison, Jane. *Meander, Spiral, Explode: Design and Pattern in Narrative*.
   Catapult, 2019.
Anthropy, Anna. *Rise of the Videogame Zinesters: How Freaks, Normals,
   Amateurs, Artists, Dreamers, Drop-outs, Queers, Housewives, and People Like
   You Are Taking Back an Art Form*. Seven Stories Press, 2012.
Ashwell, Sam Kabo. "Standard Patterns in Choice-Based Games." *These
   Heterogenous Tasks*. January 26, 2015. heterogenoustasks.wordpress.com
   /2015/01/26/standard-patterns-in-choice-based-games/.
Bartle, Richard. "Hearts, Clubs, Diamonds, Spades: Players Who Suit MUDS."
   Journal of MUD Research 1 (1996): 19.
Baxter, Charles. *Burning Down the House: Essays on Fiction*. Graywolf Press,
   1997.
Bogost, Ian. *Persuasive Games: The Expressive Power of Videogames*. MIT Press,
   2007.
Bogost, Ian. "The Rhetoric of Video Games." In *The Ecology of Games:
   Connecting Youth, Games, and Learning*, edited by Katie Salen. The John
   D. and Catherine T. MacArthur Foundation Series on Digital Media and
   Learning, 117–40. MIT Press, 2008.
Butler, Octavia. *Kindred*. Doubleday, 1979.
Case, Julialicia. "Braving the Controller: Charting the Narrative Strategies of
   Video Games." *AWP Chronicle*. October/November 2017.
Castro, Joy. "Racial and Ethnic Justice in the Creative Writing Course." *Gulf
   Coast*. Fall 2015.
Chavez, Felicia Rose. *The Anti-Racist Writing Workshop: How to De-colonize the
   Creative Classroom*. Haymarket Books, 2021.
Cole, Alyana and Jessica Zammit. *Cooperative Gaming: Diversity in the Games
   Industry and How to Cultivate Inclusion*. CRC Press, 2020.
Dickens, Charles. *Great Expectations*. Chapman and Hall, 1861.
Freire, Paulo. *Pedagogy of the Oppressed*. Continuum, 1970.

Fullerton, Tracy. *Game Design Workshop: A Playcentric Approach to Creating Innovative Games*. A K Peters/CRC Press; 4th ed., 2018.

Gardner, John. *The Art of Fiction: Notes on Craft for Young Writers*. Vintage, 1991.

Gee, James Paul. *What Video Games Have to Teach Us About Learning and Literacy*. St. Martins Griffin, 2003.

Gibson, William. *Neuromancer*. Ace, 1984.

Gygax, Gary and Dave Arneson. *Dungeons and Dragons*. Tactical Studies Rules, 1974.

Hemingway, Ernest. *Death in the Afternoon*. Scribner's, 1932.

Hemingway, Ernest. *Men without Women*. Scribner's, 1927.

Hergenrader, Trent. *Collaborative Worldbuilding for Writers and Gamers*. Bloomsbury Academic, 2018.

Huizinga, Johan. *Homo Ludens*. Angelica Press, 2016.

Hunter, Georgia. *We Were the Lucky Ones*. Viking, 2017.

James, Marlon. *Black Leopard, Red Wolf*. Riverhead Books, 2020.

Jayanth, Meg. "Don't Be a Hero: *80 Days*—The Game." The Lit Platform. https://theliteraryplatform.com/news/2014/07/dont-be-a-hero-80-days-the -game/.

Jayanth, Meg. "Leading Players Astray: *80 Days* and Unexpected Stories." Game Developers Conference. San Francisco. March 2–6, 2015.

Jayanth, Meg. *You Can Panic Now*. blog, defunct, 2014.

Jenkins, Henry. *Convergence Culture: Where Old and New Media Collide*. New York University Press, 2008.

Lerman, Liz. *Critical Response Process: Getting Useful Feedback on Anything You Make, from Dance to Dessert*. Liz Lerman Dance Exchange, 2003.

MacDonald, Keza. "Is the Video Games Industry Finally Reckoning with Sexism?" *The Guardian*. July 22, 2020.

Machado, Carmen Maria. *Her Body and Other Parties*. Graywolf Press, 2017.

Macklin, Colleen. *Games, Design and Play: A Detailed Approach to Iterative Game Design*. Addison-Wesley Professional, 2016.

Martin, George R. R. Martin. *Game of Thrones*. Bantam Books, 1996.

McGonagal, Jane. *Reality Is Broken: Why Games Make Us Better and How They Can Change the World*. Penguin Books, 2010.

McKee, Robert. *Story*. ReganBooks, 1997.

Miller, J. Hillis. *Ariadne's Thread*. Yale University Press, 1995.

Morrison, Toni. *The Bluest Eye*. Holt, Rinehart and Winston, 1970.

Morwood, Claire. *Tutorial*. https://www.shimmerwitch.space/bitsyTutorial .html.

Nagamatsu, Sequoia. *How High We Go in the Dark*. William Morrow, 2022.

Percy, Ben. *Thrill Me: Essays on Fiction*. Graywolf, 2016.

Phillips, Amanda. *Gamer Trouble: Feminist Confrontations in Digital Culture*. New York University Press, 2020.

Poe, Edgar Allen. "The Tell-Tale Heart." *The Pioneer*. January, 1843.

Quantic Foundry. "The 9 Quantic Gamer Types." https://quanticfoundry.com/gamer-types/.

Salesses, Matthew. *Craft in the Real World: Rethinking Fiction Writing and Workshopping*. Catapult, 2021.

Sandel, Caetlyn. "Rise of the Diary Game." AlterConf. Boston, 2014. https://alterconf.com/speakers/caelyn-sandel

Schwartz, Mimi. *Writing True: The Art and Craft of Creative Nonfiction*. Cengage Learning, 2006.

Shaw, Adrienne. *Gaming at the Edge: Sexuality and Gender at the Margins of Gamer Culture*. University of Minnesota Press, 2015.

Shawl, Nisi and Cynthia Ward. *Writing the Other: A Practical Approach*. Aqueduct Press, 2005.

Shiro, Masamune. *Ghost in the Shell*. Dark Horse Comics, 1991.

Smith, Zadie. *White Teeth*. Vintage, 2001.

Stern, Jerome. *Making Shapely Fiction*. W.W. Norton, 1991.

Tekinbas, Katie Salen and Eric Zimmerman. *Rules of Play: Game Design Fundamentals*. MIT Press, 2003.

Thomsen, Michael. "Why Is the Game Industry So Burdened with Crunch? It Stars with Labor Laws." *The Washington Post*. March 24, 2021.

Tolkien, J. R. R. *The Lord of the Rings: The Fellowship of the Ring*. William Morrow, 1954.

Twain, Mark. *Adventures of Huckleberry Finn*. Chatto and Windus, 1884.

Verne, Jules. *Around the World in 80 Days*. Pierre-Jules Hetzel, 1873.

Walker, John. "Wot I Think: Papers, Please." *Rock Paper Shotgun*. August 12, 2013. www.rockpapershotgun.com/wot-i-think-papers-please.

Wallace, David Foster. *Infinite Jest*. Little, Brown and Co., 1996.

Whitehead, Colson. *Zone One*. Doubleday, 2011.

Games/Media

80 Days. Inkle, 2014.

Analogue: A Hate Story. Christine Love, Love Conquers All Games, 2012.

Angry Birds. Rovio Entertainment, 2009.

Assassin's Creed. Ubisoft, 2007.

Assassin's Creed Odyssey. Ubisoft, 2018.

Assassin's Creed Origins. 2017.

Asteroids. Atari, 1979.

Avengers: Endgame. Walt Disney, 2019.

Back to the Future. Universal Pictures, 1985.

Baldur's Gate. BioWare, 1998.

Batman: Arkham VR. Rocksteady Studios, 2016.

Bertie the Brain. Josef Kates, 1950.

Bioshock. 2K, 2007.

Blade Runner. Warner Brothers, 1982.

Blood and Truth. London Studio, 2019.

Braid. Jonathan Blow, 2008.

Bravely Default. Silicon Studio, 2012

Breath of Fire. Capcom, 1993.

Brothers: A Tale of Two Sons. Starbreeze Studios, 2013.

Butterfly Soup. Brianna Lei, 2017.

Call of Duty. Infinity Ward, 2003.

Cave Story. Studio Pixel, 2004.

Celeste. Maddy Makes Games, 2018.

Chrono Cross. Square, 1999.

Chrono Trigger. Square, 1995.

Cibele. Star Maid Games, 2015.

Citizen Sleeper. Jump Over the Age, 2022.

Colossal Cave Adventure. William Crowther, 1976.

Coming Out on Top. Obscurasoft, 2014.

Crash Bandicoot. Naughty Dog, 1996.

Crosscode. Radical Fish Games, 2018.

Cyberpunk 2077. CD Projekt Red, 2020.

Crystal Castles. Atari, 1983.

Danganronpa. Spike Chunsoft, 2010.

Darkest Dungeon. Red Hook Studios, 2016.

Darkwood. Acid Wizard Studio, 2017.

Dead Cells. Motion Twin/Playdigious, 2018.

Dead Space. Electronic Arts, 2008.

Dear Esther. The Chinese Room, 2012.

Depression Quest. Zoe Quinn, 2013.

Detroit: Become Human. Quantic Dream, 2018.

Deus Ex. Ion Storm, 2000.

Disco Elysium. ZA/UM, 2019.

Dishonored 2. Arkane Studios, 2016.

Doki Doki Literature Club. Team Salvato, 2017.

Donkey Kong. Nintendo, 1981.

Don't Starve. Klei Entertainment, 2013.

Doom. id Software, 1993.

Dragon Age: Origins. BioWare, 2009.

Dragon Quest. Square Enix, 1986.

DragonBall Z. Fuji TV, 1989.

Driver. Reflections Interactive, 1999.

Dys4ia. Anna Anthropy, Newgrounds, 2012.

Earthbound. Nintendo, 1994.

Elden Ring. FromSoftware, 2022.

The Elder Scrolls V: Skyrim. Bethesda Game Studios, 2011.

Escape From New York. AVCO Embassy Pictures, 1981.

Eternal Darkness. Silicon Knights, 2002.

Every Day the Same Dream. Paolo Pedercini, 2009.

Everybody's Gone to the Rapture. The Chinese Room, 2016.

Everything. David OReilly, Double Fine Productions, 2017.

Everything Everywhere All at Once. A24, 2022.

Fallout 4. Bethesda Game Studios, 2015.

Final Fantasy VI. Square, 1994.

Final Fantasy VII. Square, 1997.

Final Fantasy IX. Square, 2000.

Fire Emblem: Three Houses. Intelligent Systems, 2019.

Firewatch. Campo Santo, 2016.

The Fiscal Ship. Serious Games Initiative, 2018.

Florence. Ken Wong, Mountains, 2018.

Foldit. U of Washington Center for Game Science, 2008.

Freshman Year. Nina Freeman, 2015.

Furious 7. Universal Pictures, 2015.

Game of Thrones. HBO, 2011. (TV Series)

Gato Roboto. Doinksoft, 2019.

Gauntlet. Atari Games, 1985.

Ghost of Tsushima. Sucker Punch, 2020.

Giant Bomb. Jeff Gerstmann, Fandom Inc,. 2008 to present. giantbomb.com

Goat Simulator. Coffee Stain Studios, 2014.

The Godfather. Paramount Pictures, 1972.

Gone Home. Fullbright Company, 2013.

Grand Theft Auto. Rockstar North, 1997.

Grand Theft Auto III. Rockstar North, 2001.

Grand Theft Auto: Vice City. Rockstar North, 2002.

Grand Theft Auto: San Andreas. Rockstar North, 2004.

Hades. Supergiant Games, 2020.

Half-Life. Valve, 1998.

Half-Life: Alyx. Valve, 2020.

Her Story. Sam Barlow, 2015.

The House in Fata Morgana. NOVECT, 2012.

Horizon: Forbidden West. Guerilla Games, 2022.

Horizon: Zero Dawn. Guerilla Games, 2017.

Illusion of Gaia. Quintet, 1993.

The Incredibles. Pixar, 2004.

Jurassic Park. Universal Pictures, 1993.

Kimmy. Star Maid Games, 2017.

King's Quest. Sierra On-Line, 1984.

Lady Bird. A24, 2017.

The Last of Us. Naughty Dog, 2013.

The Last of Us II. Naughty Dog, 2020.

The Legend of Zelda. Nintendo, 1987.

The Legend of Zelda II: The Adventure of Link. Nintendo, 1987.

The Legend of Zelda: Breath of the Wild. Nintendo, 2017.

Life Is Strange. Dontnod Entertainment, 2015.

Life Is Strange: True Colors. Deck Nine, Square Enix, 2021.

Lifeline. 3 Minute Games, 2015.

Liyla and the Shadows of War. Rasheed Abueideh, 2016.

lonelygirl15. Beckett, Flinders, Goodfried, and Goodfried, 2006.

Mafia. Illusion Software, 2002.

Mafia II. 2K Czech, 2010.

Marvel's Spider-Man. Insomniac Games, 2018.

Mass Effect. BioWare, 2017.

The Matrix. Warner Brothers, 1999.

Maze War. NASA, 1973.

McDonald's Videogame. Molleindustria, 2006.

Metal Gear Solid. Konami, 1998.

Metal Gear Solid IV. Konami, 2008.

Metal Gear Solid V: The Phantom Pain. Konami, 2015.

Minari. A24, 2020.

Minecraft. Mojang Studios, 2011.

Mobile Suit Gundam. Nippon Sunrise, 1979. (TV Series)

Monument Valley. Ustwo Games, 2014.

Moving Out. DevM Games, 2020.

Need for Speed III: Hot Pursuit. Electronic Artists, 1998.

Neon Genesis Evangelion. Gainax Tatsunoko, 1995. (TV Series)

Never Alone. E-Line Media, 2014.

The New Adventures of Peter and Wendy. Shawn DeLoache and Kyle Waters, 2014.

NieR:Automata. PlatinumGames, 2017.

Night in the Woods. Infinite Fall/Secret Lab, 2017.

Noby Noby Boy. Namco Bandai Games, 2009.

No Man's Sky. Hello Games, 2016.

No Man's Sky VR. Hello Games, 2019.

No More Heroes. Goichi Suda, 2007.

Octodad: Dadliest Catch. Young Horses, 2014.

Omega Virus. Michael Gray, Milton Bradley, 1992.

Outlast. Red Barrels, 2013.

Oxenfree. Night School Studio, 2017.

Papers, Please. Lucas Pope, 3909 LLC, 2013.

Pentiment. Obsidian Entertainment, 2022.

Pier Solar. WaterMelon, 2010.

Pitfall!. Activision, 1982.

Pong. Atari, 1972.

Portal. Valve, 2007.

The Portopia Serial Murder Case. Chunsoft, 1983,

Prince of Persia. 1989, Broderbund.

Queers in Love at the End of the World. Anna Anthropy, 2016.

Raiders of the Lost Ark. Paramount Pictures, 1981.

Ratchet and Clank: Rift Apart. Insomniac Games, 2021.

Red Dead Redemption. Rockstar Games, 2010.

Red Dead Redemption 2. Rockstar Games, 2018.

Rise of the Tomb Raider. Crystal Dynamics, 2015.

Sayonara Wild Hearts. Simogo, 2019.

Shadow of the Tomb Raider. Eidos-Montreal, 2018.

She's Gotta Have It. Focus Features, 1986.

Shovel Knight. Yacht Club Games, 2014.

Slacker. Orion, 1990.

Snatcher. Konami, 1988.

Sonic the Hedgehog. Sega, 1991.

Space Invaders. Taito/Atari, 1978.

Spacewar! Steve Russell, 1962.

Spec Ops: The Line. YAGER, 2012.

Spent. Urban Ministries of Durham, 2012.

Spider-Man: Miles Morales. Sony Interactive Entertainment, 2020.

Spyro the Dragon. Insomniac Games, 1998.

The Stanley Parable. Galactic Café, 2011.

The Stanley Parable: Ultra Deluxe. Galactic Café, 2022.

Star Trek: The Original Series. NBC, 1966.

Star Wars: Episode IV—A New Hope. 20th Century-Fox, 1977.

Star Wars: Episode I—The Phantom Menace. 20th Century-Fox, 1999.

Star Wars Jedi: Fallen Order. Respawn Entertainment, 2019.

Star Wars: Knights of the Old Republic. BioWare, 2003.

Stardew Valley. ConcernedApe, 2016.

Stray. Blue Twelve Studio, 2022.

Subnautica. Unknown Worlds Entertainment, 2018.

Superbetter. Games for Change, 2012.

Superhot. Superhot Team, 2016.

Super Mario 64. Nintendo, 1996.

Super Mario Bros. Nintendo, 1985.

Super Mario Odyssey. Nintendo, 2017.

Super Meat Boy. Team Meat, 2010.

Tales of Symphonia. Namco Tales Studio, 2003.

Tass Times in Tonetown. Interplay, 1986.

Tomb Raider. Core Design, 1996.

Ultima 1: The First Age of Darkness. Origin Systems 1981.

Uncharted 2: Among Thieves. Naughty Dog, 2009.

Uncharted 4: A Thief's End. Naughty Dog, 2016.

Uncharted: Legacy of Thieves. Naughty Dog, 2022.

Undertale. Toby Fox, 2015.

Unpacking. Witch Beam, 2021.

VA-11 Hall-A: Cyberpunk Bartender Action. Sukeban Games, 2016.

The Vale: Shadow of the Crown. Epic Games, 2021.

Valiant Hearts: The Great War. Ubisoft, 2014.

Vivant Ludi. Caelyn Sandel, 2014.

What Remains of Edith Finch. Giant Sparrow, 2017.

The Witcher 3: Wild Hunt. CD Projekt RED, 2015.

Wizardry. Sir-Tech, 1981.

World of Warcraft. Blizzard Entertainment, 2004.

The Writer Will Do Something. Tom Bissell and Matthew Burns, 2015.

Xenoblade Chronicles. Monolith Soft, 2010.

Xenogears. Squaresoft, 1998.

Yakuza: Like a Dragon. Sega, 2020.

Zork. Infocom, 1977.

About the Authors

Julialicia Case is an assistant professor of English and humanities at the University of Wisconsin, Green Bay, where she helped design the game writing track for the Writing and Applied Arts BFA program and is a codirector for the Center for Games and Interactive Media. She teaches numerous courses focused on game design such as digital storytelling, interactive fiction, alternate reality games, and collaborative worldbuilding, as well as courses in fiction, creative nonfiction, and literature. She earned her MA from the University of California, Davis, and her PhD from the University of Cincinnati, and she is an editor for the Digital, Multimodal, and Multimedia section of the *Journal of Creative Writing Studies*. Her fiction and creative nonfiction have appeared in *The Gettysburg Review*, *Crazyhorse*, *Witness*, *Blackbird*, and other journals. You can learn more about her writing and scholarship at www.julialiciacase.com.

Eric Freeze is the author of two short story collections: *Dominant Traits* and *Invisible Men*, as well as two books of creative nonfiction, *Hemingway on a Bike* and *French Dive*. He holds a PhD in creative writing from Ohio University, and he is an associate professor at Wabash College where he teaches fiction, creative nonfiction, screenwriting, and writing for video games. He is cofounder of the Wabash Game Lab, an instructional and social space for the Wabash Community to connect through gaming. His work appears in numerous periodicals, including *The Southern Review*, *Boston Review*, and *Harvard Review*. He is married to academic and birth educator Rixa Freeze, and he is the father of four bilingual soccer-crazed children. He lives half the year in Crawfordsville, Indiana, and the other half in Nice, France. More about Eric Freeze at www.ericfreeze.com

Salvatore Pane is the writer of two novels, *Last Call in the City of Bridges* and *The Theory of Almost Everything*, a book of nonfiction, *Mega Man 3*, and a story collection, *The Neorealist in Winter*. He's the winner of the 2022 Autumn House Fiction Prize, and his short work has appeared in *Indiana Review*, *American Short Fiction*, and *Story Magazine*. He is an associate professor at the University of St. Thomas, who has been teaching video game courses since 2016, and he holds an MFA from the University of Pittsburgh. He's the writer/narrative designer of *RetroMania Wrestling*, and he served as Freelance Localization Editor for NIS America's *Trails to Azure*. He plays video games daily, and his favorite titles include *Final Fantasy VI*, *Metal Gear Solid*, *Night in the Woods*, *Nier: Automata*, and *Cibele*. He lives in St. Paul, Minnesota, with his wife and Martin Scorsese the Cat and can be reached at www.salvatore-pane.com.

Index